Studies in Fuzziness and Soft Computing

Volume 438

The series "Studies in Fuzziness and Soft Computing" contains publications on various topics in the area of soft computing, which include fuzzy sets, rough sets, neural networks, evolutionary computation, probabilistic and evidential reasoning, multi-valued logic, and related fields. The publications within "Studies in Fuzziness and Soft Computing" are primarily monographs and edited volumes. They cover significant recent developments in the field, both of a foundational and applicable character. An important feature of the series is its short publication time and world-wide distribution. This permits a rapid and broad dissemination of research results.

Indexed by SCOPUS, DBLP, WTI Frankfurt eG, zbMATH, SCImago.

All books published in the series are submitted for consideration in Web of Science.

Kamal Abdulla · Rafik Aziz Aliev

The Book of Dede Korkut and Fuzzy Logic

Springer

Kamal Abdulla
Azerbaijan University of Languages
Baku, Azerbaijan

Rafik Aziz Aliev
Georgia State University
Atlanta, USA

Azerbaijan State Oil and Industry
University
Baku, Azerbaijan

ISSN 1434-9922 ISSN 1860-0808 (electronic)
Studies in Fuzziness and Soft Computing
ISBN 978-3-032-05905-5 ISBN 978-3-032-05906-2 (eBook)
https://doi.org/10.1007/978-3-032-05906-2

This Springer imprint is published by the registered company Springer Nature Switzerland AG
The registered company address is: Gewerbestrasse 11, 6330 Cham, Switzerland

Dedicated to the memory of Prof. Lotfi Zadeh—Father of Fuzzy Logic

Foreword

There are great books that are turning points in the history of mankind. Although I am of the opinion that the book you have in your hands is one of such landmarks, it is a difficult undertaking to say or write anything about it. Another difficulty is that the comments about such great books often remain dull because the greatness of the book cannot be grasped. But I have a sense of comfort in the knowledge that whatever I say will be incomplete in comparison to this book.

A very old story is being told. An elephant is brought at dusk to a country that has never seen an elephant, placed in a dark tent, and the people who enter are asked to describe the elephant as they touch it. One said it looked like a long-pointed horn, another said it was like a pillar, another said it was like a sail, and another said there was no doubt it was like a trunk.

Anyone who reads this multifaceted and multidisciplinary book will have their own interpretation and conclusion. This is consistent with the democratic, tolerant, and creative identity of the authors. After discussing the book's foundations and authors, I will share my personal views on the book and its implications for me.

Obviously, this will be an interpretation and will only express my point of view. I think the most important feature that makes great books great is their openness to interpretation. Each reader of this book will focus on different aspects of it, and this will ultimately make it greater.

Not every book is suitable for everyone.

Before I begin, I would like to present my view on whether or not this book is suitable for those who pick it up. For whom is this book not suitable? For whom is it crucial, a "must-read"?

This book is not intended for individuals who hold a view of the world as being black and white, and devoid of meaning beyond mundane, mechanical routines. They should not waste their time.

This book is essential for those who believe that there is more to the world than what meets the eye, that there are unknowns beneath the obvious, that these dimensions can be explored and created by the human mind, that there are infinite options other than right and wrong, that we live in a colorful, multidimensional matrix, and who dedicate their lives to expanding the creative horizons of human civilization beyond mechanical necessities, in whatever field. They should read it absolutely and carefully. It will be worth every moment.

We are more accustomed to the concept of living systems in the natural sciences; the survival, transformation and adaptation of a plant or animal species; ecology, evolution, natural equilibrium, natural selection and elimination; the extinction of one species, the survival and continued activity of another.

Excluding structuralist studies, the presence and discussion of systems in the linguistic and social sciences can be difficult to comprehend However, when examining the definitions of the concept of system, it becomes clear that issues related to human beings and society fall within this field.

How does a work like 'The Book of Dede Korkut', which has survived since an unknown era of history, first in oral culture and then in multiple transcriptions, while generations, societies, and states have disappeared, continue to endure? What sets it apart from others when thousands of similar works have been forgotten, destroyed, or lost? How does it persist in the geographies it has passed through, in the societies and communities it has lived in, and remain effective and influential in a significant part of the world? What are its secrets? What can we learn from it about living, surviving, and succeeding?

What is the basic structure of this book? How has it maintained its structure despite all the years it has passed and the age it has reached? What is the logic, the language, the features that dominate this book and what is the importance of this language in the life of the book? How important is the linguistic structure of the book, or the logic that governs the book in general, for the book's overall longevity? Is its openness to interpretation or its flexibility important to its adaptation and survival like other natural systems?

The debate over whether man or machine will ultimately be the winner is one of the most challenging questions of our time, not only in science and technology circles, but for almost all intellectuals. The development of artificial intelligence studies and the extraordinary successes of its applications are increasingly intensifying and fueling the debate on this issue.

The question above is speculative and subjective. It depends on individuals' knowledge, thoughts, beliefs, and expectations. The real question is how to harmonize these two fields. Can humans maximize their contribution to the development of civilization, humanity, nature, and the universe while pushing the limits of themselves and the machine?

This brings us to the issue of improving human behavior. Language is undoubtedly the most important tool at humanity's disposal for this purpose. For thousands of years, our ancestors have used language to influence our behavior, as seen in the 'Book of Dede Korkut'.

Language and improving human behavior, as well as the study of artificial intelligence, have gradually become some of the most focused areas of research for humanity, especially after the Second World War, and it is possible to say that the transitivity between theory and practice, science and technology, academia and the market has increased since then.

The two areas we are talking about, artificial intelligence and human behavior improvement, are exactly the areas where theory and practice go hand in hand, where you cannot practice without theory, but theories that cannot be applied are of no value.

There is a close relationship between efforts to improve human behavior and artificial intelligence and machine learning. Both of these fields are sub-fields of cybernetics, and studies are carried out with a systems approach.

Fuzzy logic is the fundamental theory of artificial intelligence. We can make a claim on fuzzy logic because its founder, Lotfi Zadeh, is rooted in our culture. Lotfi Zadeh is an American scientist of Azeri origin. Fuzzy Logic studies are considered to have started with an article he wrote in 1965. It would not be an overstatement to say that Rafiq Aliev is the intellectual successor of Lotfi Zadeh. We can best learn about the relationship between artificial intelligence and fuzzy logic from him. However, the other area that interests us is the relationship between human behavior and fuzzy logic.

At this point, we should immediately state that Lotfi Zadeh, in his work on artificial intelligence, believed that machines that think and behave like humans cannot be made in a 1-0, right—wrong, black and white logic, because the most important feature of humans is the ability to make more real decisions with insufficient information in a fuzzy environment, and that it is human reasoning with language that makes this possible. Language itself is fuzzy and has the potential to encompass and define the entire universe. He believed that a machine that thinks like a human should not work with mathematical ambiguities, but with algorithms that use the possibilities of natural language. All he wanted was to apply the principles of fuzzy logic to social phenomena. This work is the greatest tribute to his memory. Lotfi Zadeh is part Azerbaijani and part American. I believe that it is a special feature that this unique application of his work in the field of human sciences is published in English with the signatures of his two Azerbaijani brothers.

This book is a lighthouse for those who want to understand what drives behavior, to step into the matrix, to solve the mystery of discourse. The two great guides that form the basis of this book are the Book of Dede Korkut and Lotfi Zadeh. In fact, it is an attempt to solve the mysteries of the Book of Dede Korkut with the key provided by Lotfi Zadeh. But to me it is more than that. Anyone who is in the effort to observe, understand, change and improve behaviors well knows that nothing is as it seems. While behaviors seem clear, when you start to analyze the mind behind them, you are confused. You are comfortable on the surface, but when you dive deeper you get lost. The words are in a language you speak. But your understanding of the meaning is an interpretation and you can never be sure of its accuracy.

Once inside the matrix, the most important question and problem is that you do not know whether you are inside or outside. In airplanes, there is a very simple but vital device: The gyroscope. Its only function is to show the sky and the ground. This book gives you a very important device to understand where you are. The logic on the surface of the matrix is black and white. It is shallow, it has no depth, it requires no interpretation. Inside the matrix, everything is fuzzy. It requires implication, interpretation. The logic there is different. This book will provide you a life-saving device for you on your journey through life.

In concluding, I would like to emphasize one more very crucial point. This work is a cornerstone. But it never represents an end, but a beginning. I believe that those who follow the path paved by this work will unite human intelligence and artificial intelligence at an interface in the depths of the human mind and the system.

"The Book of Dede Korkut" and "Fuzzy Logic" brims with new concept and ideas, on my opinion it is first step of application fuzzy logic in systemic form to social sciences, literature and humanitarian fields. It is hard to find adequate words to describe the fundamental importance of this volume which opens the door profound changes in the methodology of deep investigation epos, poems, and other literature texts.

Istanbul, Turkey Dr. R. Burak Eke

Preface

When Jourdan, one of Moliere's well-known characters, learned that he had been speaking in prose all his entire life, he was surprised and questioned people around him Prose?! "What is this?!"

Even today, when a modern person turns to the samples of cultural and literary heritage, created by his ancestors, he gets acquainted with the signs of overwhelming fuzzy logic thought in the depth of their meaning. And when these traces bring him to the essence of many issues that he faces in his life, he queries himself; "Fuzzy logic? What's this?".

In truth, we would not be wrong to argue that fuzzy thinking style has left its deep traces in our modern language, style of thinking and worldview. We express a wide variety of means of expression of vagueness in our daily speech. Our entire speech seems to be wrapped around by hazy, imprecise words, idioms, and grammatical elements, like snake's surround Laocoon and his boys. Looking closely, we can find signs of the existence of such a way of thinking even in the most ordinary words.

For instance, it is common for us to use semantic patterns in our everyday speech that are instantly familiar to all of us, such as "you know well," "today is humid," and "the tall guy over there." However, upon a deeper look, we will find that there isn't a clear semantic boundary between the terms "humidity," "well," and "tall" in these examples. Any speaker who uses these terms is free to understand and apply his own interpretations. Therefore, a "good" grade for some individuals may be a test score of 80% or above, while a "good" result for others may be as low as 50%.

Let's consider how the word "tall" is interpreted. In the ancient literary and cultural epic of the Azerbaijani people "The Book of Dede Korkut" and the primary research object of our book, a female figure "Boyu Uzun Burla Khatun", in consequence of her name, is regarded taller than the other Oghuz women. However, this height criterion might be ambivalent when it comes to modern women or women from other nearby countries.

Numerous expressions, such as "humid, rainy day," also have varying interpretations among individuals. These are fuzzy and imprecise terms.

Naturally, rain is a natural occurrence, but as it might appear as a drizzle, light drizzle, or downpour, it can be challenging to visualize what rain actually looks like.

Therefore, we should see the word "rain" as ambiguous, even strange, as it does not accurately and adequately represent the phenomena of rain, which manifests itself in multiple ways.

We are compelled to use vague, ambivalent concepts of the same semantic class, in order to recognize, perceive and classify such ambiguous words in the objective world. To identify this ambiguity and determine its place in our way of thinking, new approach concepts must be introduced.

Accordingly, in 1965 the fuzzy sets and the idea of fuzzy logic based on it, created by Prof. Lutfizadeh, aroused from this need.

Zadeh expected and wished that his idea would be applied to social sciences, philosophy, economics, psychology, and linguistics, which are imprecise sciences, rather than to the fields of exact sciences and technologies. He believed that the new theory of fuzzy sets would be used in the fields as humanities, philology and social sciences. In this sense, we would like to remember one point. In 1994, during an interview with the magazine "Azerbaijan International" Prof. Lutfizadeh was asked:

- How and where will fuzzy logic be used in your opinion?

"In many, many fields. I expected people in the social sciences-economics, psychology, philosophy, linguistics, politics, sociology, religion and numerous other areas to pick up on it. It's been somewhat of a mystery to me why even to this day, so few social scientists have discovered how useful it could be.

Our contribution "The Book of Dede Korkut and Fuzzy Logic" includes the marriage of mathematical and linguistic methods for providing a formalism to deal with the linguistic vagueness of evaluative expressions in literature, prose and other humanitarian fields.

Our findings provide the basis for building an ontology of investigation literature and other different humanitarian fields by using fuzzy sets and fuzzy logic.

Since the phenomena of fuzziness in the real world is also closely related with human beings, this book gives a new insight of the vast contribution of fuzzy sets in human-oriented social sciences, literature and art.

Unlike most papers and books in the topic of social field "Dede Korkut and Fuzzy Logic" provides in our opinion, first systematic, yet practical guide for researcher wishing to combine fuzzy set with literature, including epos, tale, novels etc.

The book, introduced to the reader, proves that Lutfizadeh's dream has partially come true. This work is the first attempt to apply fuzzy logic to the analysis of an ancient epic. And that is considered in this context.

Based on fuzzy logic principles, "The Book of Dede Korkut" epic can be viewed as an ancient spiritual encyclopedia of the Azerbaijani people.

We believe that the review of "The Book of Dede Korkut" epic, an ancient spiritual encyclopedia of the Azerbaijani people today on the basis of fuzzy logic principles, is also suitable for applying this theory to other examples of folklore and it offers appropriate standards and a methodological arsenal of criteria.

At the same time, we also believe that this book has highlighted the fact that linguistics is still waiting for its research in terms of fuzziness.

Applying the elements of fuzzy logic to the depths of an ancient epic, we saw "the clash of meanings" as German philosopher Schelling mentioned. We have witnessed the stages of prehistoric people's progressive approach to human value like democracy. Thus, we believe that implementing fuzzy logic on a literary sample, which has a long history, will open up new and fascinating study directions for logic and informatics specialists as well as epic scholars.

Chapter 1 is devoted to preliminary information on Zadeh's fuzzy logic. For the present book to be self-containing, elementary information on fuzzy sets, fuzzy logic and fuzzy mathematics which are used as the formal basis of the suggested study are given.

In Chap. 2 the reader is provided with all the required facts he is familiar and unfamiliar about "The Book of Dede Korkut", which occupies one of the unique and high places in the epic world. Interestingly, comparing the ancient epics of the world in the light of fuzzy logic, such as Homer's classical epics, proves how the study is effective.

Chapter 3 of "The Book of Dede Korkut" is called "fuzzy logic elements". This article discusses the uncertainties, hesitancies, and "ifs" that arise in people's minds and how fuzzy thinking develops over time. Concurrently, details regarding the foundations (facets) of fuzzy logic are presented in this chapter. In the epic text, these aspects are examined through examples.

Chapter 4 in the book is titled Building Material of Fuzziness in "The Book of Dede Korkut". This chapter mentions a range of linguistic resources that are crucial to the development of fuzziness and hedges. Pronouns, double words, numbers, and other language features that function as hedges in the epic are explored separately.

"The text of "The Book of Dede Korkut" and the persistent uncertainty of fuzzy logic" are the topics of the book's Chap. 5. It is shown that in any area where doubts arise, the reader should look for indications of fuzzy logic. As a result, the presence of such a debatable point is described as a "suspicious constant" and serves as an indication of fuzzy logic.

Chapter 6 in the book "The Book of Dede Korkut" text and the search for the third with fuzzy logic. Here, the authors operate within the fuzzy logic's broad logical column. In addition to identifying two opposing poles, they search the epic text for instances of additional situations that already exist. Studies reveal that our epic has third and even fourth situations between two opposite poles, confounding Aristotle's logic. The writers highlight this as proof of a democratic outlook in the perspective of our ancient ancestors' world.

"You are more beautiful than the most beautiful, my darling!" is the title of Chap. 7. Here, the authors depart from the predetermined fuzzy logic column and provide a collection of instances drawn from the ancient epic that are based on it.

Chapter 8 is called "The Book of Dede Korkut" is fuzzy logic in the context of epistemology. The authors' performance in this instance comes from the fuzzy logic's epistemological column. The proportion of information about the object in the story that is revealed and kept hidden is examined.

In Chap. 9 we are going to speak about the manifestation of the derivatives of the fourth facet of fuzzy logic in the epos "The Book of Dede Korkut". Being the facet of fuzzy relations, the fourth facet of fuzzy logic embraces the fuzzy relations more precisely, fuzzy independencies between the cause and effect. The foundation of this facet are fuzzy expressions (variables) and the relations based on the scheme "If....then..."

And finally, in "Epilogue" it is discussed how and in what way the scientific-methodical platform of the authors can take place during the application of fuzzy logic to the examples of the future humanitarian fields.

We believe the writers have achieved one of their main goals through accomplishing nine chapters of the book. This book appeared as the first pioneering work in the direction of applying Zadeh's theory to any other humanitarian-oriented literary-artistic example.

We anticipate that this work will be interesting not only for "Kitabi-dade Gorgud" researchers, but also for logicians, information theorists, artificial intelligence specialists, and so on. It is the high time for professionals of widely disparate fields to come together, "talk and understand," each other's and this is a crucial issue for contemporary science. We are confident that the book will attract attention as a useful example in this direction.

Of course, we hope that the work will be beneficial to college and university students, educators, and specialists in a variety of humanities subjects.

The authors of this book dedicate it to the bright memory of the renowned scientist, Prof. L. Zadeh.

Baku, Azerbaijan — Kamal Abdulla
Atlanta, USA/Baku, Azerbaijan — Rafik Aziz Aliev

Contents

Chapter 1
Fuzzy Sets and Fuzzy Logic

Fuzzy sets are a further development of the mathematical concept of a set. Sets were first studied formally by the Georg Cantor. Fuzzy sets was introduced by Lotfi A. Zadeh in 1965. Fuzzy set theory generalizes classical set theory to allow partial membership in the sense that the membership degree of an object to a set is not restricted to first the integers 0 and 1, but may take on any value in the interval [0,1]. A set in classical set theory always has a sharp boundary because membership in a set is a black-and-white concept, i.e., an object either completely belongs to the set or does not belong to the set at all.

Even though some sets do have sharp boundaries (e.g., the set of married people), many others do not have sharp boundaries (e.g., the set of happily married couples, the set of good graduates, the set of tall persons, etc.). Fuzzy set theory directly addresses this limitation by allowing membership in a set to be a matter of degree. The degree of membership in a set is expressed by a number between 0 and 1;0 means entirely not in the set, 1 means completely in the set, and a number in between means partially in the set [1].

Fuzzy Logic has many facets. The principal facets are: fuzzy set—theoretical, logical, epistemic and relational.

1.1 Facet of Fuzzy Logic

1.1.1 Fuzzy Sets [1, 2]

More often than not, the classes of objects encountered in the real physical world do not have precisely defined criteria of membership. For example, the class of animals clearly includes dogs, horses, etc. as its members, and clearly excludes such objects

K. Abdulla and R. A. Aliev, *The Book of Dede Korkut and Fuzzy Logic*, Studies in Fuzziness and Soft Computing 438, https://doi.org/10.1007/978-3-032-05906-2_1

as rocks, fluids, plants, etc. However, such objects as starfish, bacteria, etc. have an ambiguous status with respect to the class of animals. The same kind of ambiguity arises in the case of a number such as 10 in relation to the "class" of all real numbers which are much greater than 1.

Clearly, the "class of all real numbers which are much greater than 1," or "the class of beautiful women," or "the class of tall men," do not constitute classes or sets in the usual mathematical sense of these terms. Yet, the fact remains that such imprecisely defined "classes" play an important role in human thinking, particularly in the domains of pattern recognition, communication of information, and abstraction [1]. Now let us clarify this description with the aid of formulas.

Let X be a classical set of objects, called the universe, whose generic elements are denoted x. Membership in a classical subset A of X is often viewed as a characteristic function μ_A from X to {0,1} such that

$$\mu_A(x) = \begin{cases} 1 \text{ iff } x \in A \\ 0 \text{ iff } x \notin A \end{cases}$$

where {0,1} is called a valuation set; 1 indicates membership while 0—non-membership.

If the valuation set is allowed to be in the real interval [0,1], then A is called a fuzzy set [3–8]. $\mu_A(x)$ is the grade of membership of x in A.

$$\mu_A : X \to [0, 1].$$

As closer the value of $\mu_A(x)$ is to 1, so much x belongs to A.

A is completely characterized by the set of pair.

$$A = \{(x, \mu_A(x)),\ x \in X\}.$$

Here $\mu_A(x)$ denotes membership of generic element x to fuzzy set A.

Example 1 A is a fuzzy set and is deseribed as

$$\begin{aligned} A = {} & 0.1/1 + 0.3/2 + 0.5/3 + 0.7/4 + 0.8/5 + 0.9/6 \\ & + 0.95/7 + 1.0/8 + 0.97/9 + +0.9/10 + 0.8/11 \\ & + 0.7/12 + 0.5/13 + 0.3/14 + 0.1/15. \end{aligned}$$

A classical version of this fuzzy set is

A = {1, 2, 3, 4, 5, 6, 7, 8, 9, 10, 11, 12, 13, 14, 15}

Here 1, 2, 3, …, 15 are generic elements of the set A.

For graphical representation of the fuzzy set A see Fig. 1.1.

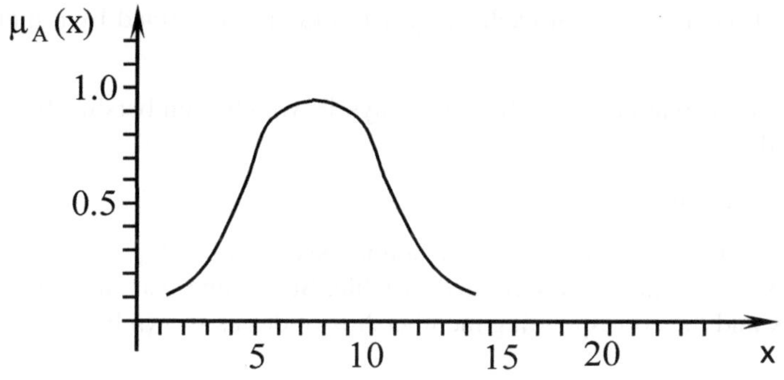

Fig. 1.1 Fuzzy set

Example 2 Assume that we describe room temperature within $[T_1, T_2]$.

The representation of temperature within a range $[T_1, T_2]$ by fuzzy and crisp sets is shown in Fig. 1.2a, b, respectively. In the first case we use membership function $[T_1, T_2] \rightarrow [0, 1]$ for describing linguistic concepts "*cold*", "*normal*", "*warm*". In the second case right—open intervals are used for describing of traditional variable by crisp sets.

Thus, it is obvious that the transitions between descriptions in fuzzy sets are considered not sharply, but regularly and gradually.

Representations and constructing of fuzzy sets

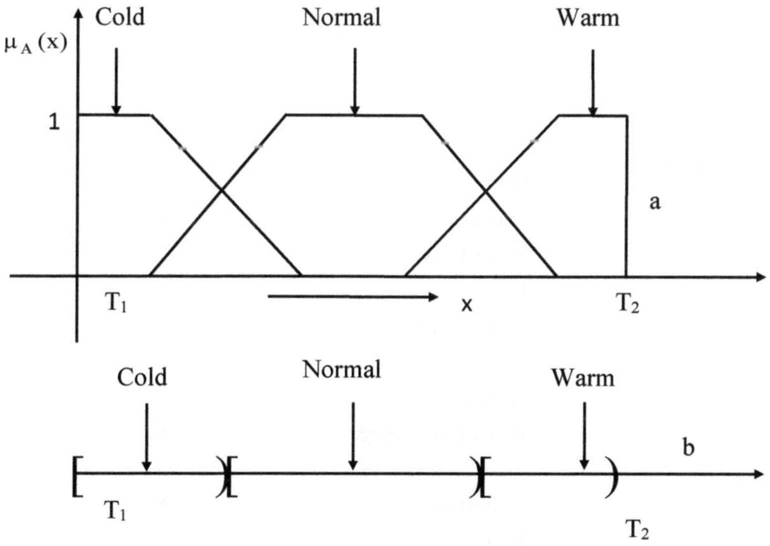

Fig. 1.2 Representation of temperature by fuzzy (Fig. 1.2a) and crisp (Fig. 1.2b) sets

It was mentioned above that each fuzzy set is uniquely defined by a membership function.

In the literature one can find different ways in which membership functions are represented.

List representation

If universal set $X = \{x_1, x_2, \ldots, x_n\}$ is a finite set, membership function of a fuzzy set A on X $\mu_A(x)$ can be represented as table. Such table lists all elements in the universe X and the corresponding membership grades as shown below

$$A = \mu_A(x_1)/x_1 + \cdots + \mu_A(x_n)/x_n = \sum_{i=1}^{n} \mu_A(x_i)/x_i.$$

Here symbol/(slash) does not denote division, it is used for correspondence between an element in the universe X (after slash) and its membership grade in the fuzzy set A (before slash). The symbol + connects the elements (does not denote summation).

Example 3 An operator in an enterprise wants to classify the quality of fruits production. The values at the variable "Quality" and elements of the universe $X = \{0, 1, 2, \ldots, 10\}$ are given in Table 1.1.

Fuzzy set $A \cong$ highest from the Table 1.1 may be represented as

$$\begin{aligned} A \cong \text{highest} = {} & 0.1/0 + 0.1/1 + 0.2/2 + 0.3/3 \\ & + 0.4/4 + 0.5/5 + 0.6/6 + 0.7/7 \\ & + 0.8/8 + 0.9/9 + 1.0/10. \end{aligned}$$

Table. 1.1 The values of the variable "quality"

Quality	$x \in X$
Very low	0
Almost low	1
Low	2
Better than low	3
Almost average	4
Average	5
Better than average	6
	7
Almost high	8
High	9
Highest	10

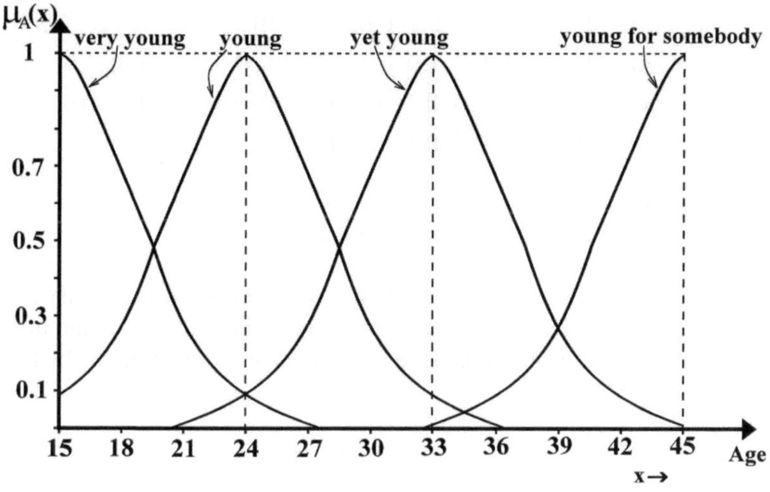

Fig. 1.3 Graphical representation of membership function

Graphical representation

Graphical description of a fuzzy set A on the universe X is suitable in case when X is one or two-dimensional Euclidean space. Simple typical shapes of membership functions are usually used in fuzzy set theory and practice.

Example 4 Assume, for instance, a group of women aged near 30. In this case we may define intuitively fuzzy sets A_1 = "*very young*", A_2 = "*young*", A_3 = "*yet young*", A_4 = "*young for somebody*" as shown in Fig. 1.3.

The numerical representation of above mentioned will be:

For comparison tabular representations of same fuzzy sets A_i, $i = \overline{1,4}$ are given below

$$
\begin{aligned}
A_1 &\cong \{1.0/15 + 0.7/18 + 0.3/21 + 0.1/24 + 0.0/27 \\
&\quad + 0.0/30 + \cdots + 0.0/45\} \\
A_2 &\cong \{0.1/15 + 0.3/18 + 0.7/21 + 1.0/24 + 0.7/27 \\
&\quad + 0.3/30 + \cdots + 0.0/45\} \\
A_3 &\cong \{0.0/15 + \cdots + 0.0/21 + 0.1/21 + 0.1/24 + 0.3/27 + 0.7/30 \\
&\quad + 1.0/33 + 0.7/36 + 0.3/39 + 0.1/42 + 0.0/45\} \\
A_4 &\cong \{0.0/15 + \cdots + 0.0/33 + 0.1/36 + 0.3/39 + 0.7/42 + 1.0/45\}
\end{aligned}
$$

where the numbers before the slashes are values of the corresponding membership function; the numbers after the slashes are arguments of the corresponding membership function which makes up a discrete universe, i.e.

$$\mu_{A_1}(x) = \{1.0, 0.7, 0.3, 0.1, 0.0, \ldots, 0.0\}$$
$$\mu_{A_2}(x) = \{0.1, 0.3, 0.7, 1.0, 0.7, 0.3, 0.1, 0.0, \ldots, 0.0\}$$
$$\mu_{A_3}(x) = \{0.0, \ldots, 0.0, 0.1, 0.3, 0.7, 1.0, 0.7, 0.3, 0.1, 0.0\}$$
$$\mu_{A_4}(x) = \{0.0, \ldots, 0.0, 0.1, 0.3, 0.7, 1.0\}$$

and the age intervals,
$X = \{15, 18, 31, 24, 27, 30, 33, 36, 39, 42, 45\}$.

Example 5 [9]. Classifying houses problem—A realtor wants to classify the house he offers to his clients. One indicator of comfort of these houses is the number of bedrooms in them. Let the available types of houses of represented by the following set.

$$U = \{1, 2, 3, 4, 5, 6, 7, 8, 9, 10\}$$

The houses in this set are described by μ number of bedrooms in a house. The realtor wants to describe a "*comfortable house for a 4-person family,*" using a fuzzy set. The fuzzy set to describe "*comfortable house for a 4-person family,*" may be defined in the following manner: *comfortable house for a 4-person family* = 0.2/1 + 0.5/2 + 0.8/3 + ¼ + 0.7/5 + 0.3/6, which means that the level of comfort for a 1- bedroom house is 0.2, a 2-bedroom house is 0.5 and so on.

The problem of constructing membership functions is problem of knowledge engineering [2].

There are many methods for estimation of membership functions. They can be classified as follows:

1. Membership functions based on heuristics.
2. Membership functions based on reliability concepts with respect to the particular problem.
3. Membership functions based on more theoretical demand.
4. Membership functions as a model for human concepts.
5. Neural networks based construction of membership functions.

The heuristic methods for estimation of the membership use partial information about it, such as the values that membership function takes at a finite number of samples in the universe of discourse.

For example, in order to build the membership function of the fuzzy set A = "young" or "old" the following formula is used:

$$\mu_{young}(x) = \begin{cases} 0 & \text{if } x \le 25 \\ 1/(1 + ((x - 25)/5)^{-2}) & \text{if } x > 25 \end{cases}$$

$$\mu_{old}(x) = \begin{cases} 0 & \text{if } x < 50 \\ 1/(1 + ((x - 50)/5)^{-2}) & \text{if } x \ge 50 \end{cases}$$

For example if the hero of "The book of Dede Korkud" is age 30, degree of hes youngness is calculated as

$$\mu_{young}^{Moniko}(30) = J/((1 + 1/(30 - 25)/5)^2 = 1/(1 + 1) = 0.5$$

In addition, it is necessary to perform mathematical operations on fuzzy sets.

Operations on fuzzy sets.

There exist mainly two standard fuzzy operations: fuzzy intersection and union which are generalization of the corresponding classical set operations.

Let's A and B be two fuzzy sets in X with the membership functions μ_A and μ_B, respectively. Then the operations of intersection, union and complement are defined as given below.

Fuzzy standard intersection and union [2].

The intersection (∩) and union (∪) f fuzzy sets A and B can be calculated by the following formulas:

$$\forall x \in X \ \mu_{A \cap B}(x) = \min(\mu_A(x), \mu_B(x))$$
$$\forall x \in X \ \mu_{A \cup B}(x) = \max(\mu_A(x), \mu_B(x)),$$

where $\mu_{A \cap B}(x)$ and $\mu_{A \cup B}(x)$ are the membership functions of $A \cap B$ and $A \cup B$, respectively.

Intersection operation of two sets shows to what degree the element is in both sets?

In classical set theory, an intersection between two sets contains the elements shared by these sets.

For example, the intersection of the set of tall men and the set of fat men is the area where these sets overlap.

It represents all those elements in the universe X that simultaneously reside in (or belongs to) both sets A and B, this operation is called the logical AND.

In fuzzy sets, an element may partly belong to both sets with different memberships.

A fuzzy intersection is the lower membership in both sets of each element.

Example 6 (intersection)

$$\begin{aligned}
A &= 0.15/2 + 0.41/3 + 0.66/4 + 0.85/5 \\
&\quad + 0.97/6 + 1.0/7 + 0.9/8 + 0.6/9 \\
&\quad + 0.42/10 + 0.3/11 + 0.18/12 + 0.1/13 + 0.03/14 \\
B &= 0.05/5 + 0.1/6 + 0.16/7 + 0.25/8 + 0.35/9 \\
&\quad + 0.47/10 + 0.62/11 + 0.8/12 + 0.94/13 + 1.0/14 \\
&\quad + 0.97/15 + 0.83/16 + 0.5/17 + 0.2/18 + 0.07/19 \\
A \cap B &= 0.05/5 + 0.1/6 + 0.16/7 + 0.25/8 + 0.35/9
\end{aligned}$$

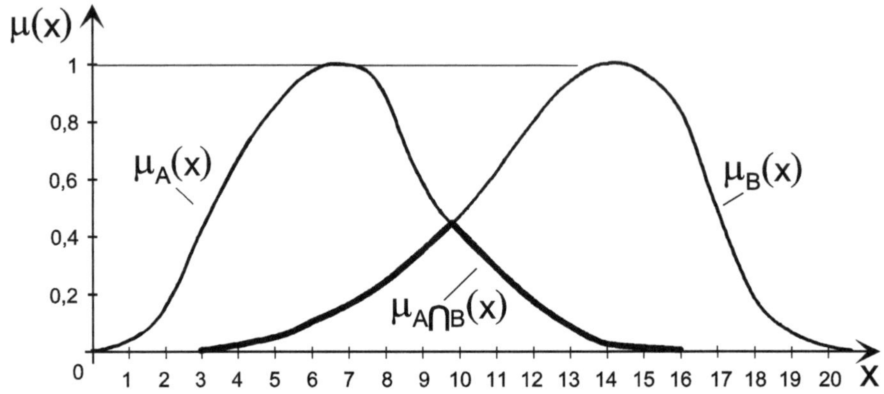

Fig. 1.4 Intersection of fuzzy sets

$$+ 0.42/10 + 0.3/11 + 0.18/12 + 0.1/13 + 0.03/14$$

For graphical representation see Fig. 1.4.

Example 7 Tall men (0/165, 0/175, 0.0/180, 0.25/182.5, 0.5/185, 1/190).

Average men (0/165, 1/175, 0.5/180, 0.25/182.5, 0.0/185, 0/190).

The intersection of these two sets is:

Tall men ∩ average men (0/165, 0/175, 0/180, 0.25/182.5, 0/185, 0/190).

Or

Tall men ∩ average men (0/180, 0.25/182.5, 0/185).

So men with 180, 182.5 and 185 are in intersection of sets Tall men and average men.

Union of fuzzy sets says: How much of the element is in either set?

In classical set, Union represents all the elements in the universe that reside in either the set A, the set B or both sets A and B. This operation is called the logical OR.

For example, the union of tall men and fat men contains all men who are tall OR fat.

In fuzzy sets, the union is the reverse of the intersection. That is, the union is the largest membership value of the element in either set.

Assume that two fuzzy sets A and B are given

$$\begin{aligned} A &= 0.07/2 + 0.2/3 + 0.4/4 + 0.63/5 + 0.87/6 \\ &\quad + 1.0/7 + 0.89/8 + 0.5/9 + +0.2/10 + 0.07/11 \\ B &= 0.05/6 + 0.11/7 + 0.21/8 + 0.32/9 + 0.46/10 \\ &\quad + 0.69/11 + 0.87/12 + +1.0/13 + 0.9/14 \\ &\quad + 0.5/15 + 0.25/16 + 0.09/18 \end{aligned}$$

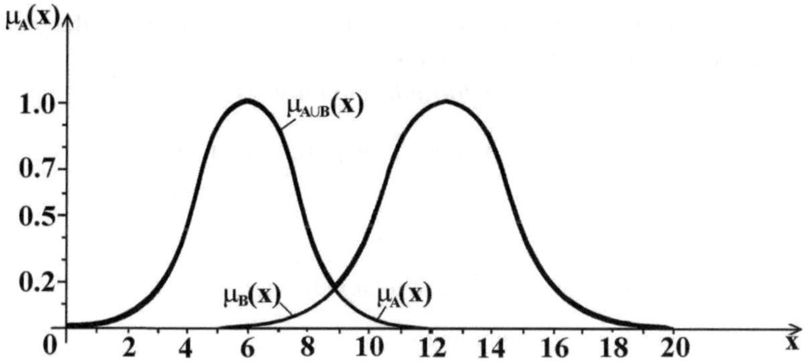

Fig. 1.5 Union of fuzzy sets

Union of A and B is

$$\begin{aligned} A \cup B = {} & 0.07/2 + 0.2/3 + 0.4/4 + 0.63/5 \\ & + 0.87/6 + 1.0/7 + 0.89/8 + 0.5/9 \\ & + 0.46/10 + 0.69/11 + 0.87/12 + 1.0/13 \\ & + 0.9/14 + 0.5/15 + 0.25/16 + 0.09/18 \end{aligned}$$

For graphical representation see Fig. 1.5.

Example 8 Tall men (0/165, 0/175, 0.0/180, 0.25/182.5, 0.5/185, 1/190)
Average men (0/165, 1/175, 0.5/180, 0.25/182.5, 0.0/185, 0/190)
The union of these two sets is:
ll men ∪ average men (0/165, 1/175, 0.5/180, 0.25/182.5, 0.5/185, 1/190).

Example 9 [9]. A four person family wants to buy a house. An indication of how comfortable they are to be is the number of bedrooms in the house. But they also want a large house. Let U = {1,2,3,4,5,6,7,8,9,10} be the set of available houses described by their numbrooms. Then the fuzzy set C (for comfortable) may be described as

$$C = \left\{ \frac{0.2}{1}, \frac{0.5}{2}, \frac{0.8}{3}, \frac{1.0}{4}, \frac{0.7}{5}, \frac{0.3}{6}, \frac{0}{7}, \frac{0}{8}, \frac{0}{9}, \frac{0}{10} \right\}$$

Let L be the fuzzy set *Large* defined as

$$L = \left\{ \frac{0}{1}, \frac{0}{2}, \frac{0.2}{3}, \frac{0.4}{4}, \frac{0.6}{5}, \frac{0.8}{6}, \frac{1}{2}, \frac{1}{8}, \frac{1}{9}, \frac{1}{10} \right\}$$

Then,

$$C \cap L = \left\{ \frac{0}{1}, \frac{0}{2}, \frac{0.2}{3}, \frac{0.4}{4}, \frac{0.6}{5}, \frac{0.3}{6}, \frac{0}{7}, \frac{0}{8}, \frac{0}{9}, \frac{0}{10} \right\}$$

To interpret this, five bedrooms is optimal, but only satisfactory to the grade 0.6. The second best solution is four bedrooms.

The union of the fuzzy sets *Comfortable* and *Large* is

$$C \cup L = \left\{ \frac{0.2}{1}, \frac{0.5}{2}, \frac{0.8}{3}, \frac{1}{4}, \frac{0.7}{5}, \frac{0.8}{6}, \frac{1}{7}, \frac{1}{8}, \frac{1}{9}, \frac{1}{10} \right\}$$

Here four bedrooms is fully satisfactory (1,0) because it is comfortable, and 7–10 bedrooms also, because that would mean a large house.

Definition. Standard fuzzy complement

The complement $\overline{A}$ of A is defined by the membership function:
$\forall x \in X\ \mu_{\overline{A}}(x) = 1 - \mu_A(x)$.

As already mentioned $\mu_A(x)$ is interpreted as the degree to which x belongs to A. Then by the definition $\mu_{\overline{A}}(x)$ can be interpreted as the degree to which x does not belong to A.

Example 10 Assume that fuzzy set A is given.

$$\begin{aligned} A = {} & 0/1 + 0.05/2 + 0.14/3 + 0.27/4 + 0.5/5 \\ & + 0.76/6 + 0.93/7 + 1.0/8 + 0.96/9 \\ & + 0.84/10 + 0.62/11 + 0.37/12 + 0.25/13 \\ & + 0.16/14 + 0.09/15 + 0.03/16 + 0/17 \end{aligned}$$

Complement of A will be

$$\begin{aligned} \overline{A} = {} & 1.0/1 + 0.95/2 + 0.86/3 + 0.73/4 + 0.5/5 \\ & + 0.24/6 + 0.07/7 + 0/8 + 0.04/9 \\ & + 0.16/10 + 0.38/11 + 0.63/12 + 0.75/13 \\ & + 0.84/14 + 0.91/15 + 0.97/16 + 1.0/17. \end{aligned}$$

Example 11 Given a fuzzy set of tall men.
Tall men (0/180, 0.25/182.5, 0.5/185, 0.75/187.5, 1/190)
the fuzzy set of NOT tall men will be:
NOT tall men (1/180, 0.75/182.5, 0.5/185, 0.25/187.5, 0/190).

Properties of Fuzzy Sets [10, 11]

If we compare the properties of crisp sets with fuzzy sets, we observed that following three properties are not similar.

- Law of Contradiction.
- Law of Excluded Middle (In this book, we will frequently encounter the third and fourth cases, which are also included in "The book of Dede Korkut". But here, the third case is not excluded).

– Absorption of Complement.

As we will move further in the chapter, we will verify these properties for both continuous and discrete fuzzy sets through some examples.

So, if we compare the crisp sets and fuzzy sets, we see that three properties are not followed by the fuzzy sets, like we have a law of contradiction. So, when we use fuzzy sets instead of crisp sets, we see that there is a contradiction and so that is why the law of contradiction comes into picture and similarly the law of excluded middle also comes.

And another one the third one the absorption of complement that is not followed by the fuzzy sets and as we move further in this lecture, we'll verify these properties for both continuous and discrete fuzzy sets through various examples.

Law of Contradiction

For a crisp set A,

$$A \cap \overline{A} = \phi$$

For a fuzzy set A,

$$A \cap \overline{A} \neq \phi$$

This is called the "Law of Contradiction".

So, let us now discuss the set theoretic properties one by one and the first one is the law of contradiction. So, let us take first crisp set and see what is happening with the crisp set if we take A ∩ ´A. So, if we take any crisp set A, $A \cap' A = \phi$.

Now, if we take the case of a fuzzy set, so let's say we have a fuzzy set *A* and we do the same intersection, we take the same intersection on this fuzzy set *A* and its complement. So, this intersection is never a null set. So, we clearly see that crisp set, there is a difference in the intersection of crisp set and its complement and then intersection between fuzzy set and its complement. And this has been clearly mentioned here that if we have a crisp set A, intersection between its complement is a null set. And if A is a fuzzy set, $A \cap' A \neq \phi$; that means, a not a null set.

So, as we are trying to verify whether $A \cap' A = \phi$ or not so, let us now first have A here and then try to find ´A. So, ´A is the complement of A. So, we see here is a complement of A fuzzy set which is represented by the green color. So, this is we can say the complement of fuzzy set A and this is this will be denoted by ´A which is complement of fuzzy set A and this is fuzzy set A fuzzy set A.

So, if we take intersection of these two fuzzy sets so, we already know if we are taking the intersection of any two fuzzy sets, we follow the criteria; we follow the condition as the min we take the minimum of the membership values of the corresponding generic variable values at each and every point. And this is needless to say that all of these generic variable values must belong to the universe of discourse.

Example 12 Consider fuzzy set, namely the Senior citizens set(B).

$$\mathrm{B} = \{(5,0),(15,0),(25,0),(35,0),(45,0),\\ (55,0.3),(65,0.9),(75,1),(85,1)\}$$

Compliment: $\mathrm{B}^{\mathrm{C}} = \{(5,1), (15,1), (25,1), (35,1), (45,1)\}$

$$\mathrm{B} \cap \mathrm{B}^{\mathrm{C}} = \{(5,0),(15,0),(25,0),(35,0),(45,0),\\ (55,0.3),(65,0.1),(75,0),(85,0)\}$$

Actually by the law of classical set contradiction, $\mathrm{B} \cap \mathrm{B}^{\mathrm{C}} = \emptyset$ (An empty set). But in the case of fuzzy sets the Axiom of Contradiction fails.

Law of Excluded Middle [12].

For a crisp set A,

$$\boldsymbol{A} \cup \overline{\boldsymbol{A}} = \boldsymbol{X}$$

For a fuzzy set A,

$$\boldsymbol{A} \cup \overline{\boldsymbol{A}} \neq \boldsymbol{X}$$

This is called the "Law of Excluded Middle".

Then we have another set theoretic property and this is called the Law of Excluded Middle.

So, if we take why is it called law of excluded middle, we will get to know in a moment. So, if we have crisp set A and if we take the $A \cup \tilde{A}$. So, for crisp sets we always get the universe of discourse as a result, whereas if we take a fuzzy set A and if we do the same operation means we take the $A \cup \tilde{A}$, A this is never going to be the universe of discourse. So, here we have clear cut difference in between the crisp set the union of crisp sets and fuzzy sets. So, we call this discrepancy as this contradiction as the law of excluded middle. So, here as I already mentioned that if we take crisp set A if we take union of $A \cup \tilde{A}$, we should get the whole universe of discourse. Whereas, if we take fuzzy sets A, $A \cup \tilde{A}$ is not going to give you the universe of discourse. So, that is what is the difference and this is called the law of excluded middle.

$$\mathrm{B} \cup \mathrm{B}^{\mathrm{C}} = \{(5,1),(15,1),(25,1),(35,1),(45,1),\\ (55,0.7),(65,0.9),(75,1),(85,1)\}$$

By the Axiom of Excluded Middle, $\mathrm{B} \cup \mathrm{B}^{\mathrm{C}} = \mathrm{X}$. But in the case of fuzzy sets the Axiom of Excluded Middle fails.

Example 13 "The book of Dede Korkut" embodies the principle of excluded third. In various stories, a third case is included between the two opposite poles. For example, between the poles of "either dead or alive", the ancient Oghuz tribes imagine a third state, called "little death". This is, in our modern medicine, state of coma. Is he dead or alive—maybe he is in a coma?!

Here we consider absorption of complement for intersection.
For crisp sets A and B,

$$A \cap (\overline{A} \cup B) = A \cap B$$

For fuzzy sets A and B,

$$A \cap (\overline{A} \cup B) \neq A \cap B$$

1.2 Fuzzy Number

A fuzzy number is 2 a fuzzy set A on R which possesses the following properties: (a) A is a normal fuzzy set; (b) A is a convex fuzzy set; (c) α-cuts of A A^{α} are a closed interval for every $\alpha \in (0,1]$; (d) the support of A A^{+0} is bounded.

In Fig. 1.6 some basic types of fuzzy numbers are shown.

Arithmetic operation on fuzzy numbers

There are different methods for developing fuzzy arithmetic [1, 2].

Let two triangular fuzzy numbers A and B be given (see Fig. 1.6a)

$$A = (a_1, a_2, a_3); B = (b_1, b_2, b_3) \tag{1.1}$$

Then we define.
Sum of A and B

$$A + B = (a_1 + b_1, a_2 + b_2, a_3 + b_3) \tag{1.2}$$

Inverse of A

$$A^{-1} = (-a_3, -a_2, -a_1) \tag{1.3}$$

Subtraction of A and B

$$A - B = (a_1 - b_3, a_2 - b_2, a_3 - b_1) \tag{1.4}$$

Multiplication of fuzzy number A to crisp member k

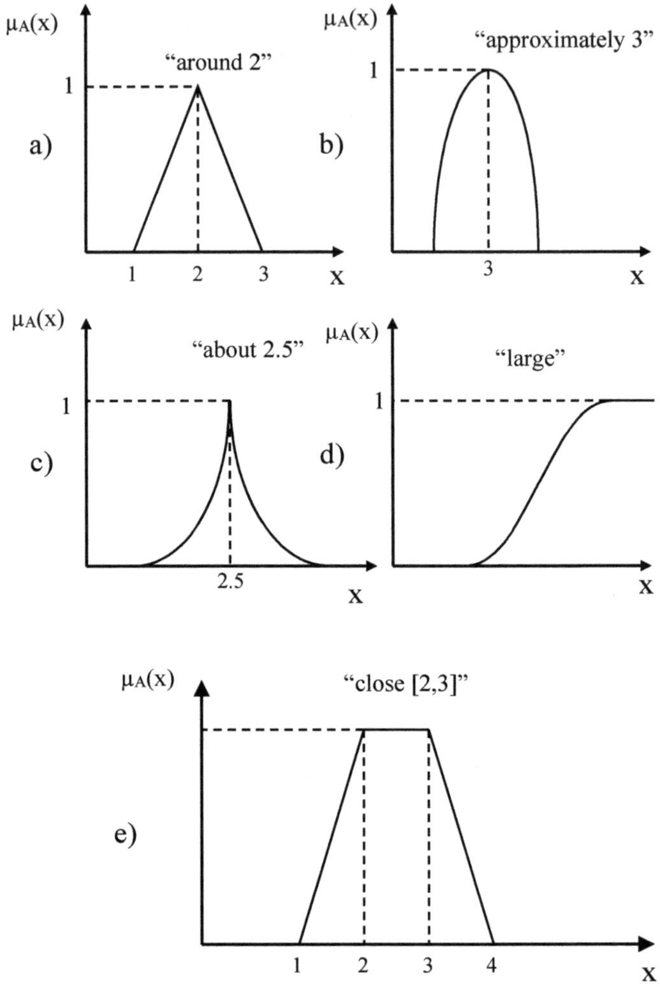

Fig. 1.6 Types of fuzzy numbers

$$k \cdot A = \begin{cases} (ka_1, ka_2, ka_3) & \text{if } k > 0 \\ (ka_3, ka_2, ka_1) & \text{if } k < 0 \\ (0, 0, 0) & \text{if } k = 0 \end{cases} \tag{1.5}$$

Produst of fuzzy numbers A and B

$$A \cdot B \approx (a_1 \cdot b_1, a_2 \cdot b_2, a_3 \cdot b_3), a_1, b_1 \geq 0 \tag{1.6}$$

Division of fuzzy sets A and B

$$A : B \approx (a_1/b_1, a_2/b_2, a_3/b_3), a_1 \geq 0, b_1 > 0,$$
$$0 \notin \text{supp } A, 0 \notin \text{supp } B. \tag{1.7}$$

Example 14 Two fuzzy number are given

$$A = (2,\ 4,\ 5), B = (4,\ 5,\ 8)$$
$$A + B = (2 + 4,\ 4 + 5,\ 5 + 8) = (6,\ 9,\ 13),$$
$$A^- = (-5,\ -4,\ -2),$$
$$A - B = (2 - 8,\ 4 - 5,\ 5 - 4) = (-6,\ -1,\ 1)$$

If $k = 3$ then

$$k \cdot A = (3 \cdot 2,\ 3 \cdot 4,\ 3 \cdot 5) = (6,\ 12,\ 15)$$
$$A \cdot B \approx (2 \cdot 4,\ 4 \cdot 5,\ 5 \cdot 8) = (8,\ 20,\ 40),$$
$$A : B \approx (2/4,\ 4/5,\ 5/8) = (1/2,\ 4/5,\ 5/8).$$

Example 15 Lets consider Aliev's gift problem. Andrew usually presents an expensive gift to Ann for her birthday [13]. They agreeed to present gift no so expensive. From now on Andrew presents gift not so expensive. What is diffence between the first and the third gift values? Let we have three linguistic terms for "gift": "not so expensive", "medium expensive", "expensive". The fuzzy sets are shown on Fig. 1.7. Lets consider difference between, "expensive" (A_1) and "not so expensive" (A_3) gift.

A difference between gift values ($A_1 - A_3$) is calculated by use of fuzzy arithmetic:

$$A_{13} = A_1 - A_3 \tag{1.8}$$

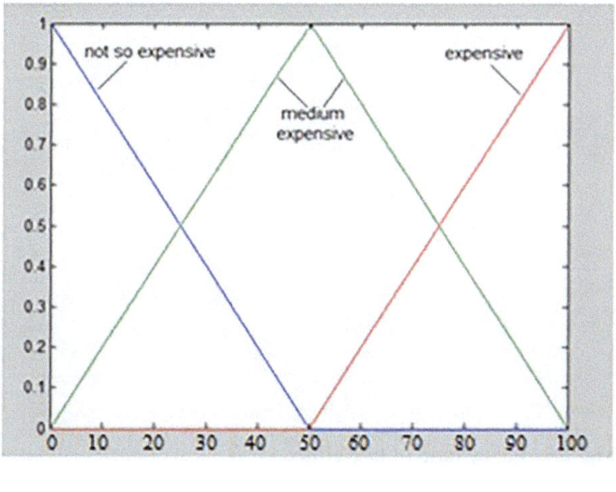

Fig. 1.7 The fuzzy sets for "gift"

So, $A_{13} = (0, 100, 100)$. (see Fig. 1..1.8) The defuzzified value of A_{13} is approx. 75. So, difference between "expensive" and "not so expensive" gifts is around 75.

The ratio $A_{31} = A_3/A_1$ approximated by triangular fuzzy number is $A_{31} = (0, 0, 1)$ (see Fig. 1.9). The defuzzified value of A_{31} is approx. 0.25. So, new "not so expensive" gift is about 4 times less expensive than "expensive" one.

***Example 16* Zadeh's two boxes problem**

Assume that we have two open boxes, A and B, each containing twenty black and white balls. A ball is picked at random. If I pick a white ball from A, I win a1 dollars; if I pick a black ball, I lose a2 dollars. Similarly, if I pick a white ball from B, I win b1 dollars; and if I pick a black ball, I lose b2 dollars. I am shown the boxes for a few

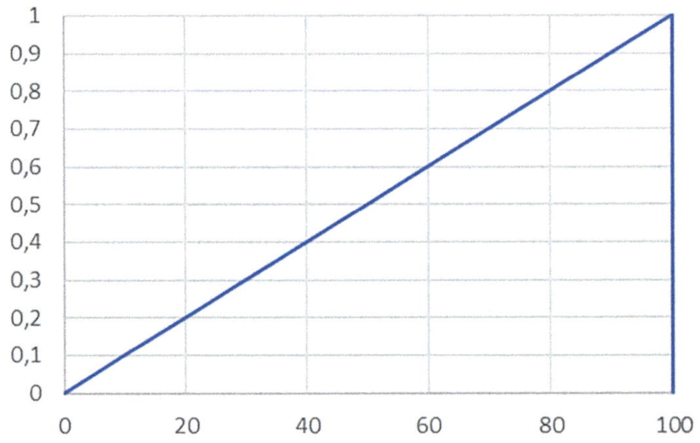

Fig. 1.8 Fuzzy difference between two gifts

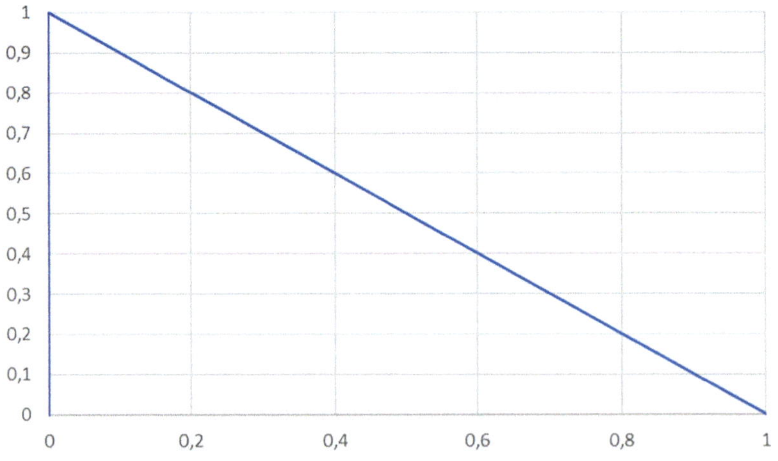

Fig. 1.9 Fuzzy ratio of two gifts

seconds, not enough to count the balls. I form a perception of the number of white and black balls in each box. At the same time, assume that I don't know the exact amounts of dollars I can gain or lose – a1, a2, b1, b2 – but only perception-based approximate values for them are provided in form of linguistic imprecise evaluations.

Given perception-based information on probabilities, gains and losses, which box should I choose?[19, 20].

Solution

Let us denote boxes as A and B and colors of balls as w (white) and b (black), respectively.

The set of possible events will be represented as $\{Aw, Bw, Ab, Bb\}$, where Aw means "a white ball picked from box A", Bb means "a black ball picked from box B".

Denote probabilities of the events as $\tilde{P}(Aw), \tilde{P}(Bw), \tilde{P}(Ab), \tilde{P}(Bb)$.

Denote the outcomes of the events as $\tilde{X}(Aw), \tilde{X}(Bw), \tilde{X}(Ab), \tilde{X}(Bb)$, where $\tilde{X}(Aw)$ denotes the outcome faced when a white ball picked from box etc.

The alternatives are: $\tilde{f}_1$ (means choosing box A), $\tilde{f}_2$ (means choosing box B).

Assume that imperfect decision-relevant information is treated by using fuzzy outcomes and fuzzy probabilities given in form of the following fuzzy numbers:

$$\tilde{X}(Aw) = \$(15, 20, 25),\ \ \tilde{X}(Ab) = \$(-10, -5, 0),$$
$$\tilde{X}(Bw) = \$(80, 100, 120),\ \ \tilde{X}(Bb) = \$(-25, -20, -15).$$
$$\tilde{P}(Aw) = (0.25, 0.35, 0.5, 0.6), \tilde{P}(Bw) = (0.1, 0.2, 0.25, 0.35).$$
$$\tilde{P}(Ab) = (0.4, 0.5, 0.65, 0.75), \tilde{P}(Bb) = (0.65, 0.75, 0.8, 0.9).$$

Now we can solve the problem by calculating fuzzy overall utilities for every boxes $\tilde{U}(\tilde{f}_1),\ \tilde{U}(\tilde{f}_2)$. The computed $\tilde{U}(\tilde{f}_1),\ \tilde{U}(\tilde{f}_2)$ shown in Fig. 1.10 [14].

Applying comparison fuzzy numbers we find that: $\tilde{U}(\tilde{f}_1) \geq \tilde{U}(\tilde{f}_2)$. So, first box will be shown.

Example 17 **Bingo game** [15]

Let a transparent Bingo Blower box contain red, blue and green balls numbers of which cannot be visually defined exactly but can only be defined by means of imprecise linguistic evaluations as follows: "the number of blue balls is medium, the number of red balls is about thirty percents and the number of green balls is the least". Let us consider two alternative games to choose from:

\$100 is given if a red ball is extracted, \$10 is given if a blue ball is extracted and \$10 is taken if a green ball is extracted;
\$10 is given if a red ball is extracted, \$10 is taken if a blue ball is extracted and \$100 is given if a green ball is extracted.

We have constructed the following TFNs-described probabilities:

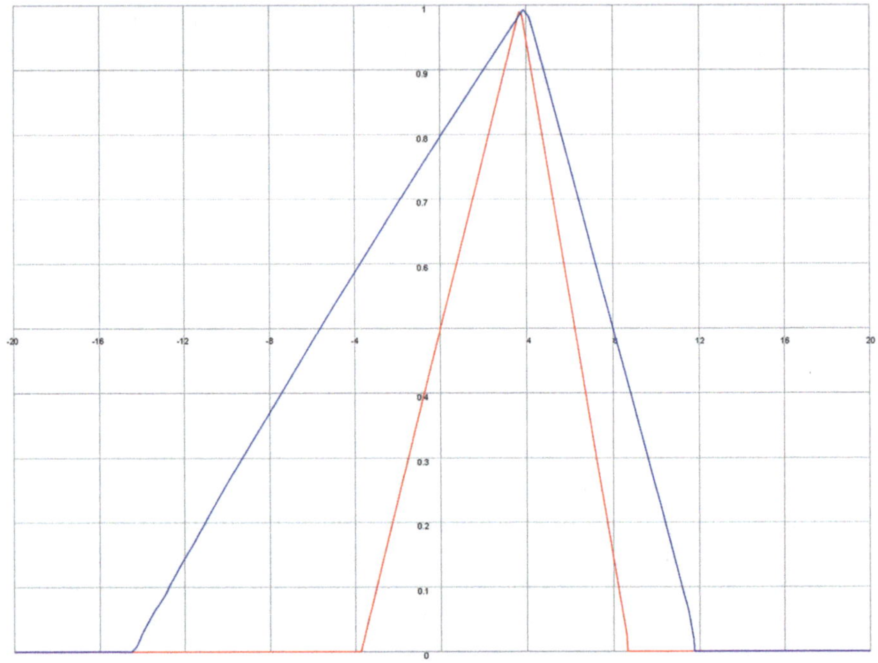

Fig. 1.10 Computed fuzzy overall utilities $\tilde{U}(\tilde{f}_1)$, $\tilde{U}(\tilde{f}_2)$

$$\tilde{P}(S_1) = (0.15, 0.2, 0.25),\ \tilde{P}(S_2) = (0.45, 0.5, 0.55),$$
$$\tilde{P}(S_3) = (0.25, 0.3, 0.35).$$

The alternatives we have are as follows:

$$f_1 = (f_1(S_1) = 100; f_1(S_2) = 10; f_1(S_3) = -10),$$
$$f_2 = (f_2(S_1) = 10; f_2(S_2) = -10; f_2(S_3) = 100).$$

The overall fuzzy utilities obtained [1]:

$$\tilde{U}(f_1) = (5.8, 8.3, 8.7),\ \tilde{U}(f_2) = (1.8, 6.15, 8.6).$$

So, 1st Alternative Game is Better.

Let a transparent Bingo Blower box contain red, blue and green balls numbers of which cannot be visually defined exactly but can only be defined by means of imprecise linguistic evaluations as follows: "the number of blue balls is medium, the number of red balls is about thirty percents and the number of green balls is the least". Let us consider two alternative games to choose from:

$100 is given if a red ball is extracted, $10 is given if a blue ball is extracted and $10 is taken if a green ball is extracted;
$10 is given if a red ball is extracted, $10 is taken if a blue ball is extracted and $100 is given if a green ball is extracted.

We have constructed the following TFNs-described probabilities:

$$\tilde{P}(S_1) = (0.15, 0.2, 0.25),\ \tilde{P}(S_2) = (0.45, 0.5, 0.55),$$
$$\tilde{P}(S_3) = (0.25, 0.3, 0.35).$$

The alternatives we have are as follows:

$$f_1 = (f_1(S_1) = 100; f_1(S_2) = 10; f_1(S_3) = -10),$$
$$f_2 = (f_2(S_1) = 10; f_2(S_2) = -10; f_2(S_3) = 100).$$

The overall fuzzy utilities obtained [1]:

$$\tilde{U}(f_1) = (5.8, 8.3, 8.7), \tilde{U}(f_2) = (1.8, 6.15, 8.6).$$

1.3 Linguistic Variable [2]

The research have shown lately that traditional analysis methods for systems analysis and computer modeling, based on precise processing of numerical data, are not capable of dealing with huge complexity of real-world problems. This leads to the fact that when solving those real-world problems we need to move away from traditional requirements for measurement accuracy which are necessary for mathematical analysis of precisely defined mechanical systems.

The necessity to sacrifice the precision and determinary is dictated also by the appearance of some classes of real-world problems, which are connected with decision-making by operator in the "man-computer" interface. Implementation of the dialog in such interface is impossible without application of languages close to natural ones and capable of describing fuzzy categories near to human notions and imaginations. In this connection, it is valuable to use the notion of linguistic variable first introduced by Zadeh [16]. Such linguistic variables allow an adequate reflection of approximate in-word descriptions of objects and phenomena in the case if there is no any precise deterministic description. It should be noted as well that many fuzzy categories described linguistically even appear to be more informative than precise descriptions.

For example, the phrase "Jone is 25 years old" can be expressed approximately by "Jone is young". In this sense, the word "young" can be considered as a linguistic value for the variable "age". Here we take into account that the linguistic value plays the same role as the numerical value 25 does. The same can be said about the linguistic values such as "very young", "a bit older than", "almost of middle age" and so forth, if we match them with numerical values 17, 24, and 31.

The set of values of a linguistic variable has made the term-set of this variable. This set in general may have unlimited number of elements though in practice it is finite. For example, a term-set of the linguistic variable "age" can be written as follows:

T(age) = "very young" + "almost young" + "young" + ⋯ + "almost of middle age" + "of average age" + ⋯ + "old" + "old enough", "very old".

Here the sign ' + ' stands for union. Note, that in case of the linguistic variable "age" a numerical variable "age", for example, taking the values of 17, 24, 31, 38,..., 94, is so-called basis variable of the linguistic variable of age. In this case such a linguistic value as "old" can be interpreted as a name for some fuzzy restriction on the values of the basis variable. This restriction essentially is the sense of the linguistic value "old".

The fuzzy restriction on the values of the basis variable is characterized by a compatibility function which maps each value of the basis variable into some number from the interval [0,1], characterising the relation of this value with the fuzzy restriction. Note that mathematically and conceptually the compatibility function is identical to the membership function introduced below.

If u is the name for a fuzzy variable, then the restriction associated with that name can be interpreted as the meaning of the fuzzy variable u. For example, if the restriction associated with the value "old" of the fuzzy variable "age" represents a fuzzy subset of the set X = [17,94] of the type:

$$\mathrm{R}(\text{“old”}) = \int_{17}^{94} \left\{1 - \left[1 - \frac{(x-17)}{77}\right]\right\} / \mathrm{x}, \quad \mathrm{x} \in \mathrm{X}.$$

Example 18 Assume that the author of "The book of Dede Korkut" is 67 years old. Then

$$\mathrm{R}^{\mathrm{Nik}}(\text{“old”}) = 1 - \left(1 - \frac{67-17}{77}\right) = 1 - (1 - 0.65)$$
$$= 1 - 0.35 = 0.65$$

then this fuzzy set can be considered as the meaning for the variable "old".

Another important aspect of the notion of linguistic variable is that a linguistic variable is associated with the two rules: the syntactic rule, which can be set as a grammar, generating names for the variable; and the semantic rule, which determines an algorithmic procedure for calculating the meaning of each value. Thus these rules make the essential part of the description of the structure of linguistic variable.

The range of possible values of a linguistic variable represents the universe of discourse of that variable.

For example, the universe of discourse of the linguistic variable speed might be the range between 0 and 220 km/h and may include such fuzzy subsets as very slow, slow, medium, fast, and very fast.

Universe: Elements of a fuzzy set are taken from a *universe of discourse* or *universe* for short. The universe contains all elements that can come into consideration. Even the universe depends on the context, as the following example shows.

Example 19 (Universe)

The set of *young people* could have all human beings in the world as its universe. Alternatively it could be the numbers between 0 and 100; these would then represent age.

The set x > 10 could have as a universe all positive measurements.

A linguistic variable carries with it the concept of fuzzy set qualifiers, called hedges.

Example 20 (linguistic variable)

Consider the sentence "The amount of trading is heavy" which uses a fuzzy set "Heavy' to describe the quantity of the stock market trading in one day. More formally, this is expressed as:

Trading Quantity is Heavy

The variable *Trading Quantity,* which is a linguistic variable, in this example demonstrates an important concept in fuzzy logic.

Hedges are terms that modify the shape of fuzzy sets. They include adverbs such as very, somewhat, quite, more or less and slightly [17].

Hedges act as operations themselves, For instance "Very" performs concentration and *creates a new subset*. From the set of tall men, it derives the subset of very tall men. Extremely serves the same purpose to a *greater extent*.

More or less performs dilation; for example, the set of more or less tall men is broader than the set of tall men (dilation is an operation opposite to concentration. It expands the set).

Indeed, the operation of intensification, it can be done by increasing the degree of membership above 0.5 and decreasing those below 0.5.

Hedges are useful as operations, but they can also break down continuums into fuzzy intervals:

For example, the following hedges could be used to describe temperature: very cold, moderately cold, slightly cold, neutral, slightly hot, moderately hot and very hot.

Hedges often used in practical applications:

Extremely, *same purpose as* very, for example: 0.86 membership in the set of tall men will be 0.6361 membership in the set of extremely tall men.

Very very, an extension of concentration, for example: 0.86 membership in the set of tall men will be 0.5470 in the set of very very tall men.

More or less, the operation of dilation *expands a set and increases the degree of membership*, for example: 0.86 membership in the set of tall men will be 0.9274 membership in the set of more or less tall men.

Indeed, the operation of intensification, it can be done by increasing the degree of membership above 0.5 and decreasing those below 0.5.For example: 0.86 membership in the set of tall men will be 0.9608 membership in the set of indeed tall men and 0.24 membership in tall men set, will have a 0.1152 membership in the indeed tall men set.

Very, *the operation of concentration*, it reduces the degree of membership of fuzzy elements using a mathematical square equation, for example: 0.86 membership in the set of tall men will be 0.7396 membership in the set of very tall men.

1.4 2nd Facet—Fuzzy Logic [1, 2]

According to one of the most widely accepted definitions, logic is an analysis of methods of reasoning. In studying these methods, logic is mainly interested in the form, not in the content of the arguments used in reasoning process. The truth or falsity of individual premises or consequence is not considered in logic. Here the main issue is establishing weather the truth of the consequence can be inferred from the truth of premises. Systematic formalization of the correct approaches of reasoning is one of the main problems of logic.

In logic it is possible to make new and more complex statements by aggregating simple statements using different approaches. Here we shall consider only the truth-functional combinations in which truth or falsity of new statements is defined by the truth or falsity of the components of the statement [1, 2].

One of the simple operations on the statements is negation.

For example, if A is a statement then its negation is denoted as $\neg$ and is read as "not A".

Another truth-valued operation on the statements A and B is conjunction AND denoted as A&B or $A \wedge B$. The statement is true if both of the statements A and B are true. The statements A and B are called conjunctive members of the conjunction A&B.

The disjunction operation on the statements A and B corresponds to the connective OR and is denoted by $A \vee B$. In natural language the connective OR is used in two different senses: separating and connecting sense.

The following important truth-valued operation is implication *If* A *Then* B. The above is false if premise A is true and consequence B is false. The implication *If* A *Then* B is denoted by $A \rightarrow B$.

Expression "A iff B" is usually denoted as A $\equiv$ B. This expression is called equivalence. It is obvious that A $\equiv$ B is true iff A and B have the same truth value.

The truth values for all these logical operations are given in Table 1.2.

Table 1.2 Truth values for logical operations

A	B	¬ A	A&B	$A \vee B$	A ⊃ B	A ≡ B
T	T	F	T	T	T	T
F	T	T	F	T	T	F
T	F	F	F	T	F	F
F	F	T	F	F	T	T

The symbols ¬, &, ∨, ⊃, ≡ are called propositional connectives. Each statement, constructed by these connectives has some truth value depending on the truth values of the statement's components. The propositional form is an expression constructed from the propositional letters (variables) A, B, C etc. by the propositional connectives.

Each propositional form defines some truth function, which may be represented as truth tables for this propositional form. Truth function of n arguments is an arbitrary function of n arguments which takes truth values T (true) or F (false) if its arguments take the same values.

A propositional form, which is true independently of the values given by propositional letters is called tautology. The propositional form is called tautology iff the corresponding truth function takes only the value T.

For example, the following logic formulas are propositional tautologies:

(1) $\neg(A\&\neg A)$ is negation of contradiction;
(2) $((A \vee B) \rightarrow (\neg A \rightarrow B))$ is expression of disjunction by negation and implications;
(3) $(((\mathrm{A} \rightarrow \mathrm{B})\&\mathrm{A}) \rightarrow \mathrm{B})$.

In its turn, propositional form which is false under all possible truth values of propositional variables is called contradiction. For example, propositional forms $(A \equiv \neg A)$ or $(A\&(\neg A))$ are contradions.

It should be noted that the implication has the following important property called "modus-ponens": IF (A → B) is true and A is true then B is true. This rule is also called the first form of hypothetical syllogism. "Syllogism" means the deductive conclusion in which one judgement is a necessary consequence of two others.

The fuzzy logic is a generalization of the classical logic and deals with the ambiguity in the logic. Here some properties of the classical logic and the fuzzy logic [18].

As in our ordinary informal language, "sentence" is used in the logic. A sentance having only "true (1)" or "false (0)" as its truth value is called "proposition".

Example 21 The following sentences are propositions.

The Oghuz hero Basat killed Tepegoz.	(true)
$2 + 4 = 7$	(false)

In the fuzzy expression (formula), a fuzzy proposition can have its truth value in the interval [0,1]. The fuzzy expression function is a mapping function from [0,1] to [0,1].

$$f : [0, 1] \rightarrow [0, 1]$$

If we generalize the domain in n-dimension, the function becomes as follows:

$$f : [0, 1]^n \rightarrow [0, 1]$$

The fuzzy logic is a logic represented by the fuzzy expression (formula) which satisfies the followings.

(i) Truth values, 0 and 1, and variable x_i ($\in[0,1]$, $i = 1, 2, \ldots, n$) are fuzzy expressions.
(ii) If f is a fuzzy expression, ~ f is also a fuzzy expression.
(iii) If f and g are fuzzy expressions, $f \wedge g$ and $f \vee g$ are also fuzzy expressions.

There are some operators in the fuzzy expression such as $\neg$ (negation), $\wedge$ (conjunction),
$\vee$ (disjunction), and $\rightarrow$ (implication).
We can see that the two properties are not satisfied in the following examples.

(a) Law of contradiction,
(b) Law of excluded middle.

1.5 3rd Facet—Epistemic Facet of Fuzzy Logic

It is taken for granted in much of the literature on vagueness that semantic and epistemic approaches to vagueness are fundamentally at odds. If we can analyze borderline cases and the sorites paradox in terms of degrees of truth, then we don't need an epistemic explanation. Conversely, if an epistemic explanation suffices, then there is no reason to depart from the familiar simplicity of classical bivalent semantics [19].

If "tall Burla khatun" has a classical extension, then there is a shortest tall man. Of course, we have no way of knowing how tall the shortest tall man is. And even if we could know, the placement of the line between the tall and the non-tall would appear arbitrary.

Epistemicism is an attempt to bite this bullet, by explaining on general epistemological grounds why we should expect to be ignorant in just this way, and by rejecting as verificationist the idea that we should be in a position to know exactly where the semantic boundaries lie. According to the epistemic approach, what distinguishes vague language from non-vague language has nothing to do with truth-conditions.

One popular alternative to classical semantics is to suppose that truth comes in degrees. The most common form of this view represents these degrees by real numbers between 0 and 1, with 1 representing complete truth, 0 complete falsity, and the intermediate values various degrees of "partial truth." The extensions of predicates are then naturally understood as fuzzy sets, or mappings from objects to degrees of truth. Thus, "tall Burla khatun" may map a 190 sm man onto 1, a 185 sm man onto 0.75, a 183 sm man onto 0.68, and so on. Small differences in height will yield small differences in the degree to which the predicate is satisfied. So as we look at shorter and shorter men, we will see a slow, steady decline in the degree to which "tall Burla khatun" applies, rather than a sudden, precipitous change from inclusion in the extension to non-inclusion.

Such a theory affords an attractive analysis of the sorites paradox. Lets dive same information about sorites paradox.

The classical semantics is not correct for vague discourse, as in this case.

The **sorites paradox** (sometimes known as the **paradox of the heap**) is a paradox that results from vague predicates. A typical formulation involves a heap of sand, from which grains are removed individually. With the assumption that removing a single grain does not cause a heap to become a non-heap, the paradox is to consider what happens when the process is repeated enough times that only one grain remains: is it still a heap? If not, when did it change from a heap to a non-heap?

The paradox is as follows: consider a heap of sand from which grains are removed individually. One might construct the argument, using premises, as follows [20]:

1,000,000 grains of sand is a heap of sand (Premise 1)
A heap of sand minus one grain is still a heap. (Premise 2)
Repeated applications of Premise 2 (each time starting with one fewer grain) eventually forces one to accept the conclusion that a heap may be composed of just one grain of sand. Read (1995) observes that "the argument is itself a heap, or sorites, of steps of *modus ponens*":
1,000,000 grains is a heap.
If 1,000,000 grains is a heap then 999,999 grains is a heap.
So 999,999 grains is a heap.
If 999,999 grains is a heap then 999,998 grains is a heap.
So 999,998 grains is a heap.
If...
... So 1 grain is a heap.

Then tension between small changes and big consequences gives rise to the sorites Paradox...There are many variations...[some of which allow] consideration of the difference between being...(a question of fact) and seeming...(a question of perception).

The removal of one drop from the ocean, will not make it 'not an ocean' (it is still an ocean), but since the volume of water in the ocean is finite, eventually, after enough removals, even a litre of water left is still an ocean.

More adequate method for explanation of this paradox is to use a fuzzy logic. In this context, the problem is with the principle of bivalence: the sand is either a heap or is not a heap, without any shades of gray.

Fuzzy logic offers a continuous spectrum of logical states represented in the unit interval of real numbers [0,1]—it is a many-valued logic with infinitely-many truth-values, and thus the sand transitions gradually from "definitely heap" to "definitely not heap", with shades in the intermediate region. Fuzzy hedges are used to divide the continuum into regions corresponding to classes like *definitely heap*, *mostly heap*, *partly heap*, *slightly heap*, and *not heap*.

Suppose we have a line of 10 men of gradually increasing height. Man 0 satisfies "tall man" to degree 0, man 1 to degree 0.01, man 2 to degree 0.02, and so on up to Man 10, who satisfies "tall man" to degree 1. Now consider the following sorites argument [19]:

Man 10 is a tall man.
If Man 10 is a tall man, Man 9 is a tall man.
If Man 9 is a tall man, Man 8 is a tall man.
:
If Man 1 is a tall man, Man 0 is a tall man.
Therefore, Man 0 is a tall man.
By using fuzzy logic analogically to preverense example this problem.

1.6 4th Facet—Relational Facet of Fuzzy Logic Fuzzy Relation [9]

A fuzzy relation generalizes the notion of a classical black-and-white relation into one that allows partial membership. For example a binary relation "friend" will classify all human relationship into either being friend or not being friend. A fuzzy relation "friend", in contrast, can describe the degree of friendship between two persons.

The classical notion of relation describes a relationship that holds between two or more objects. A relationship between two objects is represented by a *binary relation*, a relation with two arguments. For example, the parent relationship between a parent and his/her child can be represented as a binary relation. More generally, we can use an n-ary relation, a relation with n arguments, to describe a relationship between n objects. For instance, we can use an n-ary relation to describe that student X took course Y during semester Z in year W. This relation has four arguments such as took course (student, course, semester, year). An n-ary relation can be formally defined as set of ordered listed of n objects. Each list describes a case in which the relation holds.

A binary relation on variables x and y, whose domains are X and Y respectively, can be defined as a set of ordered pairs in $X \times Y$. For instance, the binary relation "less than" between two real numbers can be formally defined as

$$R = \{(x, y)\, x < y;\ x,\ y \in R\}$$

It is easy to see that the relation is a subset of X × Y. In general, an n-ary relation on $x_1, x_2, \ldots, x_n$ whose domains are $X_1, X_2, \ldots, X_n$ is a subset of $X_1 \times X_2 \times \cdots \times X_n$.

Since a relation can be viewed as a set, we can easily generalize the classical notion of relation using fuzzy sets. We will show how to generalize binary relations. First we can represent a binary relation R on x, y with domains X, Y as a function that maps an ordered pair (x, y) in X × Y to 0 (i.e.,$R = X \times Y \rightarrow \{0, 1\}$).

A fuzzy relation generalizes the classical notion of relation into a matter of degree. As mentioned earlier, the fuzzy relation *friend* describes the degree of friendship between two persons. Similarly, a fuzzy relation *Petite* between height and weight of a person describes the degree by which a person with a specific height and weight is considered petite. Formally, fuzzy relation R between variables x and y, whose domains are X and Y, respectively, is defined by a function that maps ordered pairs in X × Y to their degree in the relation, which is a number between 0 and 1, i.e., $R = X \times Y \rightarrow [0.1]$.

More generally, a fuzzy n-ary relation R in $x_1, x_2, \ldots, x_n$ whose domains are respectively, is defined by a function that maps an n-ary $(x_1, x_2, \ldots, x_n)$ in $X_1, X_2, \ldots, X_n$ to a number in the interval [0, 1]. i.e.,$R : X_1 \times X_2 \times \cdots \times X_n \rightarrow [0.1]$. Just as a classical relation can be viewed as a set, a fuzzy relation can be viewed as a fuzzy subset. From this perspective, the mapping above is equivalent to the membership function of a multidimensional fuzzy set.

If the possible values of x and y are discrete, we can express a fuzzy relation in a matrix form. For example, suppose we wish to express a fuzzy relation Petite in terms of the height and the weight of a female. Suppose the range of the height and the weight of interest to us are {152, 155, 157, 160, 163, 165, 168}, denoted h, and {90, 95, 100, 105, 110, 115, 120, 125} (in lb,) denoted w, respectively. We can express the fuzzy relation in a matrix form as shown in table 1.3.

Table. 1.3 A fuzzy relation—"Petite"

	90	95	100	105	110	115	120	125
152	1	1	1	1	1	1	0.5	0.2
155	1	1	1	1	1	0.9	0.3	0.1
157	1	1	1	1	1	0.7	0.1	0
160	1	1	1	1	0.5	0.3	0	0
163	0.8	0.6	0.4	0.2	0	0	0	0
165	0.6	0.4	0.2	0	0	0	0	0
168	0	0	0	0				
0	0	0	0					

Each entry in the matrix indicates the degree a female with the corresponding height (i.e., the row heading) and weight (i.e., the column heading) is considered to be *petite*. For instance, the entry corresponding to a height of 5′3″ and a weight of 115 lb. has a value 0.3, which is the degree to which such a female person will be considered a *petite* person; i.e., petite (5′3″, 115 lb) = 0.3.

Once we define the Petite fuzzy relation, we can answer two kinds of questions:

- What is the degree that a female with a specific height and a specific weight considered to be petite?
- What is the possibility that a petite person has a specific pair of height and weight measures?

Assume, two fuzzy sets $\mu_A(x)$ and $\mu_B(y)$ are given on crisp sets X and Y, respectively. Then, it is said, that fuzzy relation $R_{AB}(X, Y)$ is given on sets X and Y, if it is defined in the following way.

$$\mu_{R_{AB}}(x, y) = \min_{x,y}[\mu_A(x), \mu_B(y)]$$

for all pairs (x, y), where $x \in X$ and $y \in Y$. As above, fuzzy relation R_{AB} is defined on Cartesian product.

Example 22 Let $X = \{1, 2, 3\}$ and $Y = \{Q, L, M\}$. Let membership functions of fuzzy sets A of set X and B of set Y, respectively, have forms.

$$\mu_A = \{0.3/1, 0.8/2, 0.5/3\},\ \mu_B = \{0.2/Q, 0.9/L, 0.1/M\}$$

Then

$$\begin{aligned} R_{AB} = \{&0.2/(1, Q),\ 0.3/(1,L), 0.1(1,M),\\ &0.2/(2,Q), 0.8/(2,L), 0.1(2,M),\\ &0.2/(3,Q), 0.5/(3,L), 0.1\ (3,M)\}\ . \end{aligned}$$

Example. A fuzzy relations "Friend" describe the degree of friendship between two persons (in contrast to either being friend or not being friend in classical relation).

Composition of Fuzzy Relations

The composition operation combines fuzzy relations over different product spaces. This operation is basically gives us the flexibility to combine two or more relations on fuzzy relations, this can be defined in many ways as suitable; here is one general definition which is also known as min–max operation. The definition is a s follows:

$$R \circ S = \{(x,\ z) | (x, y) \in R_{X*Y}, (y, z) \in S_{Y*Z}\}$$

$$\mu_{R\bigcirc S}(x,\ y) = \max_{\forall y \in Y}\{\min\{\mu_R(x, y), \mu_S(y,\ z)\}\} :$$

$$R \in X^{\times}Y,\ S \in Y^{\times}Z.$$

Example 23 [21]

Let's take the following fruit example with the colour-maturity relation R.

R	Verdant	Half-mature	Mature
Green	1	0.2	0
Yellow	0.3	1	0.4
Red	0	0.2	1

And also define for a maturity-taste relation S.

S	Sour	Tasteless	Sweet
Verdant	1	0.2	0
Half-mature	0.7	1	0.3
Mature	0	0.7	1

Then by applying the composition operation to the elements of these two tables, the following is obtained:

$T = R \circ S$	Sour	Tasteless	Sweet
Green	1	0.2	0

Usually fuzzy relations are described by fuzzy IF…THEN rules.
Fuzzy rules:

A fuzzy rule can be defined as a conditional statement in the form [2, 17]:

IF x is A THEN y is B

where x and y are linguistic variables; and A and B are linguistic values.

Antecedent—The initial (or "if") part of a fuzzy rule.
Consequent—The final (or "then") part of a fuzzy rule.

An example might be a temperature controller:

If the break temperature is HOT
and the speed is NOT VERY FAST
then break SLIGHTLY DECREASED.

What is the difference between classical and fuzzy rules?

A classical IF–THEN rule uses binary logic, for example,

Rule: 1

IF speed is > 100
THEN stopping distance is long

Rule: 2

IF speed is < 40
THEN stopping distance is short

The variable speed can have any numerical value between 0 and 220 km/h, but the linguistic variable stopping distance can take either value long or short.

In other words, classical rules are expressed in the black-and-white language of Boolean logic.

We can also represent the stopping distance rules in a fuzzy form [17]:

Rule: 1

IF speed is fast
THEN stopping distance is long

Rule: 2

IF speed is slow
THEN stopping distance is short

In fuzzy rules, the linguistic variable speed also has the range (the universe of discourse) between 0 and 220 km/h, but this range includes fuzzy sets, such as slow, medium and fast. The universe of discourse of the linguistic variable stopping distance can be between 0 and 300 m and may include such fuzzy sets as short, medium and long.

Fuzzy rules relate fuzzy sets:

In a fuzzy system, all rules fire to some extent, i.e., fire partially. If the antecedent is true to some degree of membership, then the consequent is also true to that same degree.

These fuzzy sets provide the basis for a weight estimation model. The model is based on a relationship between a man's height and his weight:

IF height is tall THEN weight is heavy

The value of the output or a truth membership grade of the rule consequent can be estimated directly from a corresponding truth membership grade in the antecedent. This form of fuzzy inference uses a method called monotonic selection.

A fuzzy rule can have multiple antecedents, for example:

IF project_duration is long
AND project_staffing is large
AND project_funding is inadequate

THEN risk is high.
IF service is excellent
OR food is delicious
THEN tip is generous.

The consequent of a fuzzy rule can also include multiple parts, for instance [17]:

IF temperature is hot
THEN hot_water is reduced;
cold_water is increased.

References

1. Zadeh, L.A., Aliev, R.A.: Fuzzy Logic Theory and Applications, Part I and Part II. World Scientific (2019)
2. Aliev, R.A., Aliev, R.R.: Soft Computing and Its Applications. World Scientific (2001)
3. Zadeh, L.A.: Fuzzy sets. Inf. Control. **8**, 338–353 (1965)
4. Aliev, R.A., Mamedova, G.A., Aliev, R.R.: Fuzzy Sets Theory and Its Application. Tabriz University Press, Tabriz (1993)
5. Klir, G.J., Yuan, B.: *Fuzzy Sets and Fuzzy Logic: Theory and Applications*. PRT Prentice Hall, NJ (1995)
6. Zimmermann, H.J.: Fuzzy Set Theory and Its Applications. Kluwer Academic Publishers. Second Revised Edition, Boston (1993)
7. Klir, G.J., Clair, U.S., Yuan, B.: *Fuzzy Set Theory. Foundations and Applications*. PTR Prentice Hall, NJ (1997)
8. Aliev, R.A., Aliev, F.T., Babaev, M.D.: Fuzzy Process Control and Knowledge Engineering. Verlag TUV Rheinland, Koln (1991)
9. Unit 1.: Fuzzy Sets—An Introduction, Indira Gandhi National Open University (2017) https://egyankosh.ac.in/bitstream/123456789/12677/1/Unit-1.pdf
10. Verma, N.K.: Fuzzy Sets, Logic and Systems and Applications. Department of Electrical Engineering Indian Institute of Technology, Kanpur, Lecture-15 (2021)
11. Aliev, R.A.: Fundamentals of the Fuzzy Logic-Based Generalized Theory of Decisions, 340p. Springer, Berlin (2013)
12. Sailaja, B., Prasad, V.B.V.N.: Exploring the Axiom of Excluded Middle and axiom of Ontradiction in Fuzzy Sets, vol. 9 (2019). ISSN: 2249-8958
13. Zeinalova, L.M.: Identification of usual numbers. Procedia Comput. Sci. **120**, 758–765 (2017)
14. Aliev, R.A., Pedrycz, W., Huseynov, O.H.: Behavioral decision making with combined states under imperfect information. Int. J. Inf. Tech. Decis. **12**(3), 619–645 (2013)
15. Aliev, R.A., Huseynov, O.H.: Decision Theory with Imperfect Information, vol. 444. World Scientific Publishing (2014)
16. Zadeh, L.A.: Outline of a new approach to the analysis of complex system and decision processes. IEEE Trans. Syst. Man Cybern. **3**, 28–44 (1973)
17. Hamardheh, Q.: Neural Network and Fuzzy Logic, Lecture 19, Operations and Rules of Fuzzy Sets (2016)
18. Fuzzy Logic.: In: First Course on Fuzzy Theory and Applications. Advances in Soft Computing, vol. 27. Springer, Berlin (2005). https://doi.org/10.1007/3-540-32366-X_8

19. MacFarlaney, J.: Fuzzy Epistemicism. Department of Philosophy, University of California, Berkeley (2008)
20. Wikipedia: The Free Encyclopedia "Sorites paradox"
21. Samanta, D.: Fuzzy Relations, Rules and Inferences. Indian Institute Texnology Kharagpur, Chapter 2 (2018)

Chapter 2
The Epos "The Book of Dede Korkut"—What We Know and What We Don't

2.1 "The Book of Dede Korkut" as an Anachronistic Text

The Epos "The Book of Dede Korkut" is an ancient Turkic (Oghuz) literary written text. Associating its author's birth date with the period of Prophet Muhammad, on the one hand, proves itself, since the introductory part of the text emphasizes that the author, i.e. Dede Korkut was born in this very period. It is this note that starts the Epos, *«Rəsul əleyhüssəlam zamanına yaxın Bayat boyundan Qorqud Ata derlər bir ər qopdu».—"Close to the time of the Prophet, on whom be peace, there appeared in the tribe of Bayat a man called Korkut Ata."* This sentence clearly indicates the time when Dede Korkut was born.

Hence, the author of the Epos turns out to have a history of approximately 1300 years. However, when trying to clarify the formation of the Epos itself, another, more hidden side of the matter manifests itself. The text describes some events which involve Dede Korkut himself and are related to the periods going back to a much earlier time than that of the Prophet, may peace be upon him, or the time later than that period. While the storyline related to the Polyphemus of the Oghuz and the Oghuz hero Basat who overcomes him, modellizes artistically the life of human society in the period of Nature and the distressful journey of its transition from Nature to Culture, some other plotlines lead to the period of the Medieval Turkic-Byzatine confrontation, to the period of already established intertribal struggles. Yet, in all these plotlines we see Dede Korkut to be the main creator. The words (statements) uttered on his behalf in the introductory part remind even by their essence the lessons on life, the world and human relations taught to co-cavemen during the Cave period. Thus, in temporal terms, the life span of Dede Korkut himself is considerably long and even inconceivable.

Based on our statements, one can conclude that the Epos is an anachronistic text as a whole. Dede Korkut, who is introduced as its author, is not a historical personality. Realistically approached, Dede Korkut, who is presented as the author of the work, couldn't have lived in such a long span of time. In accordance with the

K. Abdulla and R. A. Aliev, *The Book of Dede Korkut and Fuzzy Logic*, Studies in Fuzziness and Soft Computing 438, https://doi.org/10.1007/978-3-032-05906-2_2

statements uttered by G. Vico, the Italian scholar in cultural studies and Schelling, the German philosopher that Homer is not a personality, but a principle, it is possible and unavoidable to characterize Dede Korkut as a principle and a moral standard. Indeed, when considered profoundly, Dede Korkut is actually a principle. A principle that united the Oghuz people in different historical periods, providing its unity around the uniform moral-ethical values. A principle which has survived since the ancient period of myth until now, and continues its life sometimes vividly, and sometimes secretly. Simply, this principle is called the principle of Dede Korkut.

Based on these principles and standards, the Oghuz society has acquired its life lessons through times and even today maintains them in the persons of modern Azerbaijanis, Turks of Türkiye, and Turkmens. Likewise, in many cases these principles have been applicable to other Turkic languages and peoples, the Kipchak and Karluk Turks have also maintained some legends about Dede Korkut in their folklore this way or another. We can observe these principles in the lives of other Turkic peoples as well. If Dede Korkut is treated not as a personality, but as a principle, the ascription of its history to a more ancient mythological period can easily be substantiated too.

A Brief Plot of Each Chapter

"The Book of Dede Korkut" is an epos of chivalry consisting of an introductory part and twelve chapters (stories). This epos describes the lifestyle, internal and external fights, survival struggles, worldviews, psychological feelings and ethical codes of the ancient Oghuz—the ancestors of modern Azerbaijanis, Turks, Turkmens, and Gagauz.

Although each chapter has its own plotline, they can be viewed as a single structure merged within a great novel on a larger semantic space. The developments in individual stories are sometimes closely, clearly bound with the preceding stories and sometimes through hidden ties. Simply, it is different Oghuz heroes that vary (with slight exceptions). Their love adventures, personal bravery, loyalty to one another, their mobilization to rescue those in hostage and in trouble, their relentless struggle against external and internal enemies constitute the main storyline (unchangeable values) of these chapters.

Introduction. In this part the author of the Epos provides the members of his community with the knowledge on life and objective reality, informs them how to act optimally in various situtions. This knowledge attracts the attention as life lessons at ethical-behaviour and legal levels. The statements like *«Çıxan can geri gəlməz» ("The soul which goes out, does not come back"), «Ölən adam dirilməz» ("The man who dies is not brought to life")* are used quite independently, and being addressed to the co-cavemen who think the reverse, these statements invite them to accept these ordinary laws of life. These statements sometimes might seem primitive to us. However, their vital importance for that ancient period is doubtless. Such kind of statements also imply indirectly which period of antiquity is described in the Epos. We can accept this period as the period when human beings lived in caves. The preliminary structural embryos can be said to have been created in this very period.

Along with this air of antiquity, it is curious that the introductory part also presents information on Dede Korkut himself, mentions when he was born. It turns out that the author of the work speaks of himself, characterizes himself. Such kind of situation appears in other chapters as well. In two chapters (in the chapter "Bamsi Beyrek of the Grey Horse" and in the chapter "How Basat Killed Goggle-eye") Dede Korkut turns into a participant of the plot within the chapter as a character. Besides this, by the end of each story Dede Korkut briefs on that chapter, gives a name to a hero in accordance with the ritual and thus, approves that person's membership to the community. Also, Dede Korkut shapes the common ethical outcome originating from the chapter.

Thus, we notice that in the Epos the author Dede Korkut and the participant of the plot Dede Korkut differ from each other.

"Boghach Khan Son of Dirse Khan"—is the title of the first chapter of the Epos. In this chapter Dirse Khan, who has neither a son nor a daughter and is being exposed to mockings by other nobles, carries out various useful actions by his wife's advice (saves the debtor from his debt, fills the one who is hungry, clothes the one who is naked…) and consequently, a son is born to him. As Dirse Khan's son grows up, he turns into a hero and Dede Korkut comes to give him a name "Boghach" (Bull-man), since he has defeated the bull.

The warriors, whom Dirse Khan is keeping next to him, feel jealous of Boghach for his father's attention and speak evil of him so that his father kills Boghach. Dirse Khan shoots his arrow and wounds his son whom he has taken out hunting. Those who are hunting think him dead. They return home. The mother who does not see her son among those who return, sets for his search, and finds her son lying, bedaubed with red blood at the foot of Kazilik Mountain. She takes the son home and hiding him from all, treats his wound with her milk and mountain flowers until it is healed.

The treacherous warriors of Boghach's father find out that he has not died of the wound and fearing Dirse Khan's revenge on them, they seize the khan and march him off as a hostage to the lands of the enemy. Boghach's mother declaims her son about it. With his warriors Boghach rescues his father from the grasp of the treacherous men, and both the father and the son finally meet and return to the land of the Oghuz safe and sound.

The following can be pointed out as typical elements of fuzzy logic manifesting themselves in the text of the chapter. When Boghach's mother finds her wounded son at the foot of the mountain, she thinks that he was wounded either by lions or tigers. However, in this text Aristotle's law of the excluded third "does not work". Boghach adds a third in between the two poles suggested by his mother through her assumption. He says to his mother that neither lions nor tigers of Kazilik Mountain are to blame. It is his father's fault. Unlike Aristotle's logic, here one can insert a third in between two poles. As we can see, the excluded third "does not work" in "The Book of Dede Korkut".

Another interesting example is associated with Dirse Khan's blaming either himself or his wife for having no child at the beginning of the chapter. His wife convinces him that there are some other points between these two poles. They have not filled the hungry, clothed the naked, and rendered their assistance to the needy.

Again these points enter inbetween the two poles as a third, fourth, fifth. This way, a wider choice (more than two) is offered to the listener, which, naturally, suggests the manifestation of a democratic mood in the mind, though in a primitive form.

The second chapter is called ***"How Salur Kazan's House was Pillaged"***. When Salur Kazan goes out hunting with the Oghuz nobles, the infidels get the news (again through the spy!) and come with big troops to pillage Kazan's encampment. They capture and take away Kazan's mother, wife and son. They loot his household and heavy treasure. The enemy wants to capture Kazan's herd of sheep, but Karajuk the shepherd does not allow them to do it, manages to protect the herd, fighting against the enemy.

Salur Kazan, who is out hunting, has a nightmare. Through his dream he becomes aware of these incidents. Without taking anyone with him, Kazan sets out back to his encampment. He sees the sight of pillage and sets forth against the enemy. On his way the shepherd Karajuk joins him.

The infidels search for Kazan's wife Lady Burla the Tall among the ladies they have captured, but fail to find her, since each lady says she is Lady Burla. In that case the infidels decide to kill Kazan and Lady Burla's son Uruz, and roast some of his flesh to feed the ladies with it. The one, who refuses to eat of this meat, is thus Lady Burla. Uruz agrees to be killed and his flesh be cooked, but won't let his mother be recognized by the enemy.

Kazan arrives and bestrides the enemy. Following Kazan's tracks, the other Oghuz nobles also arrive. *«Sayılmaqla Oğuz bəyləri tükənsə olmaz». ("Of counting the nobles of the Oghuz there could be no end".)*

The Oghuz defeat the enemy in the battlefield. Salur Kazan rescues his mother, wife and son. He returns his property and treasure.

The interesting example of fuzziness in the story is the sentence *«Sayılmaqla Oğuz bəyləri tükənsə olmaz». ("Of counting the nobles of the Oghuz there could be no end").* Such kind of generalization implies an infinite set.

The third chapter is called ***"Bamsi Beyrek of the Grey Horse"***.

The Oghuz nobles pray Allah to grant Prince Bay Büre a son and Prince Bay Bijan a daughter. With the prayers of the nobles children are born to these two nobles. Prince Bay Bijan and Prince Bay Büre declare these babies betrothed in the presence of the nobles.

Prince Bay Büre sends merchants to the faraway land of Rome in order to buy decent presents for his son before the latter is grown. Arriving in Istanbul, the merchants buy a grey horse, sea-born, a bow and mace.

Meanwhile Prince Bay Büre's son grows up to be a brave young man. On their way back home, the merchants are pillaged by robbers. One of the merchants manages to escape to the Oghuz' land and sees Beyrek resting with his warriors on a meadow. When Beyrek finds out about the incident, he decides to take away the goods and return them to the merchants. Without introducing himself Beyrek fights against the robbers and returns the goods to the merchants. For his bravery, Dede Korkut comes and gives him the name "Beyrek".

Beyrek meets the Lady Chichek and falls in love with her. A competition takes place between them. Beyrek wins the lady in shooting an arrow, horse racing and wrestling. He gets the lady's consent and they decide to get married. Prince Bay Bijan is not alive any more, therefore, the Lady Chichek's brother Crazy Karchar is supposed to give his consent for her marriage. But he won't give his consent for her marriage to any one. He kills one by one all those who asks for her hand. The real, hidden reason for this strange attitude is revealed a little later in connection with another moment.

Dede Korkut himself is sent for matchmaking. Unexpectedly, Crazy Karchar threatens and drives him away. Crazy Karchar tries to murder Dede Korkut with his sword. When he raises his sword to kill Dede Korkut, the latter recites a prayer and the former's hand remains suspended aloft being withered. Further, Crazy Karchar gives his consent for the Lady Chichek to marry Beyrek. Prince Bay Büre starts the preparations for the wedding.

The lord of the castle of Bayburt gets the news that the lady whom he was supposed to marry is given to Beyrek to a marriage. And Crazy Karchar's strange actions are revealed at that moment. The flow of information finds its natural channel.

The men of the prince of Bayburt who has become hostile to the Oghuz, kidnaps Beyrek on his wedding day. Beyrek has been kept captive in the Castle of Bayburt for 16 years. His being dead or alive is kept secret from all.

A man called Yaltajuk son of Yalanji happens to have Beyrek's shirt. Dipping that shirt in blood, he brings it as the news of Beyrek's alleged death. They want to marry Beyrek's betrothed the Lady Chichek to him. A wedding tent is set.

Beyrek gets the news through the merchants who have set forth in search of him and manages to escape the captivity with the help of the daughter of the Bayburt prince. Before his escape he promises the girl that he will return to marry her.

Beyrek arrives at his betrothed the lady Chichek's wedding in the Oghuz land. He does not introduce himself to anyone. Prior to the wedding he takes part in several contests, becomes the winner and gets from Kazan Khan the permission of visiting the girl who is getting married. When meeting the lady Chichek, he introduces himself to her and the lovers rejoin.

Beyrek assaults the castle of Bayburt together with the other Oghuz nobles. He frees his 39 fellow-warriors from captivity, returns to Oghuz having laid the Castle of Bayburt to the ground. However, Beyrek fails to keep his promise to the daughter of the Bayburt prince. It is here that the systematicity in the Epos is displayed. It is not a coincidence that of all the Oghuz warriors only Beyrek is murdered in another chapter. We would say that the mythological mindset takes its revenge on him this way. Being dispersed to different chapters, again reason and outcome bring forth the Epos as a single organism.

The most curious indicator of fuzziness in this story is the cover of Crazy Karchar's reluctance to give his consent to his sister's marriage, to be more precise, its concealment. Crazy Karchar goes even against Dede Korkut, such an outstanding figure of the Oghuz community. Hence, there is an arrangement in Crazy Karchar's brain unknown to us. He acts in accordance with this arrangement, drives away anyone who asks him for his sister's hand, no matter who that person is. The overt side of the text does not provide an explanation of it. The reason remains hidden. The decoding of the hidden reason takes place a while later when Beyrek is being married. A spy brings the news to the prince of the Castle of Bayburt, "Why do you sit there, my lord?" says the spy. 'The girl that Prince Bay Bijan was going to give to you he has given to Beyrek of the grey horse.' Thus, it turns out Prince Bay Bijan has had deals secret from the Oghuz people.

So, in other words, the (epistemic) reason for Crazy Karchar's senseless actions, which does not seem real as a whole, is the failure of Prince Bay Bijan to keep his promise made by him to the nobles and Prince Bay Büre, his unfaithfulness to his commitment and his further promise to give his daughter to the Prince of Bayburt for marriage. Hence, Crazy Karchar defends this real line of his father on the surface of the Epos in a senseless (unreal) form. Therefore, Crazy Karchar drives away anyone who asks for his sister the lady Chichek's hand. It is impossible to find some other reason for Crazy Karchar's behavior. This is how one can draw out such an explanation from the hidden substratum of the text through the epistemic facet of fuzzy logic.

The fourth chapter is called ***"How Prince Uruz Son of Prince Kazan was Taken Prisoner"***.

At a gathering Kazan Khan held with the nobles of the Oghuz, he expresses his intention to take his son Uruz and go out hunting, and show him the places where he has shot arrows, wielded his sword and fought. So, all get together and go out hunting.

The infidel gets the news through a spy. Troops march against Kazan Khan. During the battle, they take Uruz the Son of Kazan Khan away as a captive. After the battle Kazan Khan, who does not see his son Uruz about him, thinks that the latter has fled for fear to their home, in other words, he has deserted the battle field.

On his return he finds out that Uruz is not at home. Then Kazan Khan realizes that his son has been imprisoned by the enemy. Kazan Khan sets forth in search of his son. The infidels send Uruz to fight against his father. When the father and son rejoin and attack the enemy, the Oghuz warriors reach the battle field one after another. A victory is won over the enemy. All return home happily.

The text introduces the Oghuz warriors who come to help Kazan as follows, *"Bu məhəldə oğuz ərənləri bir-bir yetdi." ("At this point the Oghuz nobles arrived one after another.")*.

Further, the names of these nobles are mentioned. After mentioning some of them, the text says, *"Saydığımca oğuz bəyləri tükənsə olmaz. Qazanın bəyləri həp yetdi." ("Of counting the nobles of the Oghuz there could be no end; Kazan's nobles all came.")*.

If counted, there is no end to the Oghuz nobles who arrive one after another. All arrive. Yet, not all the names are mentioned. Laconicity is covered by the hedge «hamısı gəldi» ("All arrived"). On the one hand, crisp logic (they came one by one) and on the other, fuzzy logic (all arrived) seem to face each other here.

Another point is the intention of Kazan's son to obtain information from his father on the enemy prior to the battle. He asks whether someone will take revenge, fight, if noble warriors are killed among them.

Kazan answers him as follows, *«Oğul, min kafir öldürsən, kimsə səndən qan davalamaz.» ("My son, if you kill thousand infidels, no one will fight against you for revenge.").*

If we take this structure to its main variant, the sentence will be as follows: If you kill a thousand infidels, then nobody will take revenge on you. Here Kazan wants to indicate the great number of the infidels and making use of the hedge "thousand", he reformulates it in a fuzzy way.

The fifth chapter is called ***"Wild Dumrul Son of Dukha Koja"***.

An Oghuz warrior Wild Dumrul by the name tries to save the soul of a noble warrior, to rescue him from the grasp of Azrael. Wild Dumrul, who pronounced his willingness to fight against Azrael, asks Allah to make him see Azrael for this purpose, in order he could save the dead warrior's soul.

God does not like this arrogance of Wild Dumrul, and God orders Azrael to appear to Wild Dumrul to turn him pale and take his soul.

While Azrael is about to carry out God's order, Wild Dumrul realizes the danger awaiting him and starts to beg God to save his soul. God feels pity on him and orders Azrael to find a soul instead of his soul to free his own. Wild Dumrul first visits his father and then his mother to ask for their souls. None of them wants to give their souls to him. When he visits his wife to say good-bye to her, his wife who finds out about the problem, agrees to give him her soul.

This time Wild Dumrul begs God this way: take the souls of both of us, now that you are to; if you let a soul free, let free the souls of both of us. God likes these words and pardens the souls of both of them, but instead takes away his parents' souls.

As an example of fuzzy mindware, a fragment from Wild Dumrul's address to God attracts the attention:

Ucalardan ucasan
Kimsə bilməz necəsən.
Görklü tanrı!
Neçə cahillər səni
Gögdə arar, yerdə istər.
Sən xud möminlər könlündəsən.
(You are higher than the high,
No one knows what you are like,
Beautiful God!
Many the ignorant who look for you in the sky or seek you on earth,
But you are in the hearts of the faithful.)

God is neither in the sky nor on the earth. Once God is not in one of these two poles, where then is God?! A third case finds itself inbetween the two poles: God is in the hearts of the Faithful.

The sixth chapter is called ***"The Story of Kan Turali Son of Kanli Koja"***.

Kanli Koja looks for a girl suitable for his son Kan Turali when the latter reaches the age of marriage. Neither Kan Turali, nor his father can find such a girl in the land of the Oghuz. His father gets to Trebizond in search of a girl. He brings some news to his son about Princess Seljan, who wears yellow, and is the beauty of the beauties. Kan Turali goes to Trebizond to get Princess Seljan.

Three tests await those who want to marry Princess Seljan. The person is supposed to fight separately one after another a bull, a black camel-stallion and a raging lion and defeat them. Kan Turali defeats these beasts and returns to his Motherland with the girl, having her father's consent.

As the lovers approach the Oghuz land, they are encamped to rest in a picturesque site. Meanwhile, Princess Seljan's father repents of his consent for his daughter's marriage. He sends armed troops to chase them and return his daughter. United Princess Seljan and Kan Turali fight against these troops and defeat them.

Because of Princess Seljan's help to him, Kan Turali wants to murder her in order not to hear the words *"he was in need of a girl's help"* in the Oghuz land. They shot arrows at each other. Further, Princess Seljan convinces Kan Turali that she doesn't even think of feeling proud of her victory over the enemy. They reconcile. And Kan Turali reaches the Oghuz land together with his beloved one.

The seventh chapter is called ***"The Story of Yigenek Son of Kazilik Koja".***

Kazilik Koja, the minister of Bayandur Khan asks leave of the khan to go on a raid to Düzmürd Castle. He is captured as a prisoner and thrown into the dungeon of the castle. Kazilik Koja has remained in captivity there for 16 years.

When Kazilik Koja went to a raid, a son was born to him. He has turned into a sixteen-year-old warrior. His name is Yigenek who considers his father dead. His father's captivity is concealed from him. Yigenek, who finds out about it by coincidence, asks Bayandur Khan for permission to go for his father's rescue. Bayandur Khan lets the Oghuz nobles accompany Yigenek together with their armies. They reach Düzmürd Castle.

Prior to the battle the Oghuz warriors fight one by one with those in the castle and defeat them. Kazilik Koja is released from captivity. The Oghuz nobles go on a raid to Düzmürd Castle, and lay the castle to the grounds. Then they return to the Oghuz land.

The eighth chapter is called ***"How Basat Killed Goggle-eye".***

When everyone deserts their household and flees during the assault of the enemy on the Oghuz land, the baby son of Aruz Koja falls and is lost. A lion finds him and nurses him. After a while when everyone returns to their home, they bring news to Aruz Koja that in the thicket there is a lion which attacks the horses and climb them. He walks like a man. Aruz Koja who thinks that he is the baby boy he has lost during the time when they fled, and thus, he orders the boy be looked for and brought to his home. However, the boy cannot endure it at home and returns to the lion's den. Again Aruz Koja's people find him and lead him home. Dede Korkut comes and calls him "Basat". He says, «*Sən insansan, heyvanla yoldaşlıq etmə!*» ("*You are a human being, do not make friends with a beast!*") This way, Dede Korkut admits Basat as a member to the Oghuz community.

Aruz Koja has a shepherd. Once peris get settled to the spring. Aruz' shepherd notices them. He desires one of them and immediately violates her. When flying away, peri says, *«məndə əmanətin var, bir il sonra gəlib alarsan». ("You have left something in trust with me. When a year has passed, come and take it.").*

A year later the shepherd goes to that spring again. He finds out an awkward shape lying on the ground. From afar he throws stones at it. The shape grows as he continues beating. At this sight the shepherd gets scared and runs away.

Bayandur Khan happens to be on a hunting trip with the nobles. At the sight of this shape the nobles begin kicking it. The shape grows bigger, as they go on kicking. As soon as the shape is split down, a boy comes out of it. He has an eye on his forehead. Aruz Koja asks Bayandur Khan to give Goggle-eye to him. Bayandur khan fulfills his kind request.

At Aruz Koja's home Basat and Goggle-eye start growing together. A peri mother comes and wears a magic ring round his finger. Because of this ring, no sword can cut his body and no arrow can pierce it.

As Goggle-eye grows up, his wild passions also increase. He starts to eat human flesh. Goggle-eye blocks the flow of the river to the Oghuz land leaving the people without water. He prevents the people who want to desert the place from doing so. He climbs to the top of the mountain and imposes his terms on the Oghuz. Through Dede Korkut's mediation the people of the Oghuz agree to give Goggle-eye two men and five hundred sheep to eat every day. And two men are given to him to prepare his meals. This way the flow of the water is unblocked.

Basat goes to fight Goggle-eye. He finds out from one of the men serving Goggle-eye that his entire body is covered with armour, only inside his eye there is some flesh. Basat holds a piece of wood against the fire and then pierces it into the eye of Goggle-eye while the latter is asleep. Goggle-eye goes blind. Goggle-eye who realizes that there is no use of resisting, explains Basat how to kill him. By killing Goggle-eye, Basat rescues the Oghuz from this scourge.

The ninth chapter is called ***"The Story of Emren Son of Begil"***.

In order to defend the border on Georgia, Bayandir Khan sends Begil to the border, by giving this warrior the annual tribute coming from Georgia. Begil sets forth to guard the border.

Once Bayandir Khan invites Begil to his place. The nobles take Begil to hunting to entertain him. During the hunting Salur Kazan disappoints Begil, ascribing his skill in shooting arrows to his horse. Begil gets annoyed and returns. He decides to rebel against the Oghuz. His wife makes him change his mind and suggests him to go hunting and ease his soul. While hunting, he falls off his horse and breaks his thigh, and returning home, he takes to his bed.

The infidel's spy tells the infidel the news about Begil's being sick in bed. Being encouraged, the enemy assaults Begil's land. Whereas, in his turn, Begil's spy tells him about the assault of the enemy. Begil puts his armour on his son Emren and sends him against the enemy.

At first sight the enemy mistakes this man for Begil. As they want to flee out of fear, the scouts look closely and find out that the warrior sitting on the back of Begil's horse and wearing Begil's armour is not Begil. They approach him and when asking who he is, they find out that it is Begil's young son Emren.

The infidel prince begins his fight against Emren. When he is about to win Emren, the latter prays Allah for help. Allah bestows upon Emren the son of Begil the strength of forty men. Emren wins over the enemy.

Following the battle, Begil visits Bayandir Khan's court and reconciliation takes place.

In this chapter fuzzy logic is manifested in seeing one, but guessing another (quite an epistemic point), in imagining one through the other one's description.

The tenth chapter is called ***"Segrek Son of Ushun Koja".***

In the land of the Oghuz a man has two sons. The name of his elder son is Egrek and the younger one is Segrek. Egrek is captured and taken to Alinja Castle by the infidels, while he is resting during hunting together with his fellow warriors.

Years pass. The younger brother Segrek grows up to become a brave boy. Although his brother's captivity is being concealed from him, Segrek finds out about it and sets forth to rescue him. Salur Kazan provides him with a team of warriors.

Battles break out between the Oghuz and the enemy in front of Alinja Castle. Segrek displays a number of heroic feats and the enemy, finally, sends Segrek's captive brother Egrek to fight against him. The brothers recognize each other during the fighting and the Oghuz nobles attack Alinja Castle together and devastate it.

The eleventh chapter is called ***"How Salur Kazan was Taken Prisoner and How His Son Uruz Freed Him"***.

Kazan Khan goes out hunting. In front of Tomanin Castle he feels sleepy and falls asleep. When the Oghuz warriors fall asleep, they sleep for seven days and seven nights. This is used to be called "the little death". The infidels capture Kazan while asleep and ties him tightly and take him to the castle where they put him into a dungeon. For fear of Kazan and the Oghuz people, they want to release him from prison. They ask him to promise to praise them and never again attack them, under these conditions they will release him. Kazan Khan does not obey them and on the contrary, he swears the enemy, insults them and remains in captivity.

Kazan's son Uruz grows up to be a brave warrior. His father's captivity is kept secret from him. One day he finds out about it and wants to go to Tomanin Castle to rescue his father. The Oghuz warriors gather around him.

Seeing the Oghuz warriors in the process of preparation for the assault on Tomanin Castle, the enemy gets frightened and asks Kazan Khan to defend them against the warriors and promised to release him to go to his home.

Kazan Khan meets the Oghuz troops. He says, "Go away and let your prince come" to the warriors, who come one by one to fight against him, and so, he challenges Uruz. Uruz comes to meet his father face to face. Uruz almost overcomes his father at the fight, when Kazan Khan lets his son recognize him.

We encounter an interesting case of fuzzy logic passing through this chapter. Sleeping for seven days and seven nights would be called "the little death" by the ancient Oghuz. Kazan Khan falls into such sleep. The little death is, actually, the state of being neither alive nor dead. In modern medicine this is called "coma". Aristotle makes a logical claim that a man is either dead or alive. There is no third case here.

As an example of fuzzy logic, the words Kazan says about himself are striking, *«Min-min ərdən yağı gördümsə, oyunum dedim.»*. (*"Had I seen thousands of enemies, I said, it's fun."* If we restore the original version of the example, we get the following structure, *"If I had seen thousands of enemies, I said, it's fun."* Here the number *"thousands"* is actually used as hedge which indicates the approximate (fuzzy) number of the enemy.

The twelfth chapter is ***"The Outer Oghuz Rebelled against the Inner Oghuz and How Beyrek Died"***.

According to an ancient custom, whenever the Oghuz tribes got together, Kazan Khan would leave his encampment at the disposal of the nobles. He would take his wife Lady Burla the Tall by the hand and go out, and following this, the Oghuz people would enter the house and take away whatever they liked.

Once Salur Kazan didn't invite to this ceremony the nobles of the Outer Oghuz. Only the nobles of the Inner Oghuz pillaged his house. The nobles of the Outer Oghuz did not forgive Kazan Khan for that. They rebelled against the Inner Oghuz. They called Beyrek and wanted him to stand on their side. However, Beyrek did not turn disloyal to Kazan Khan and did not want to join the rebels. He is thus mortally wounded. Beyrek hardly manages to get home where he dies.

The nobles of the Inner Oghuz with Kazan Khan at their head decide to punish the nobles of the Outer Oghuz. The Inner Oghuz confronts the Outer Oghuz. In the battle Salur Kazan wounds Aruz Koja and dismounts him and has his brother Kara Göne cut Aruz Koja's head off. The battle ends and they reconcile.

As an example of fuzzy logic, Aruz Koja's call of the nobles of the Outer Oghuz for consultation is striking. He says, *«Əmən gəlsin, Alp Rüstəm gəlsin, Dönəbilməz Dülək Uran gəlsin, geri qalan bəylər həp gəlsin, - dedi. Dış Oğuz bəyləri həp yığnaq oldu.» ("Let Emen come, and Alp Rüstem, and Dölek Evren who knows not retreat... and all the other nobles...")* Here three nobles of the Outer Oghuz are mentioned, but all of them arrive. "All" (*həp*) is used as hedge, and denotes an infinite set.

2.2 F. H. von Diez and "The Book of Dede Korkut"

It should be emphasized that one can often encounter the similarity of plot, the state of parallelism between this ancient Oghuz epos and a number of ancient mythological texts of the peoples of the world. Among them, the ancient Greek mythological texts are the most striking mythological space. The identical plotline, the similar characteristics of the heroes are very typical for the Oghuz and Greek mythological spaces. It was the similarity (identity) of a number of storylines in the Epos with the storylines in the Greek mythological space that primarily attracted the attention of Friedrich Heinrich von Diez, the German diplomat and scholar in Eastern studies who has introduced the Oghuz epos to the world, and was first to provide the scientific community with the information about it.

Von Diez, who worked as the first Ambassador of Prussia to Türkiye in the early nineteenth century, had a deep love of Türkiye and the Turkic world. When he was working as an Ambassador in Istanbul, he obtained a lot of ancient Eastern manuscripts and on his way back from Türkiye took with him the manuscript of "The Book of Dede Korkut" alongside with numerous other manuscripts.

Von Diez attached a special attention to a chapter within this Epos. That was the chapter that tells "How Basat Killed Goggle-eye", which is a bright example of the similarity of plots with the Greek myths. It was this chapter that Diez intended to present to the scientific community. He decided to carry out the translation of this chapter into German and write his comprehensive introductory article on the Epos as a whole, and this chapter in particular. As we have stated, the main line of the chapter "How Basat Killed Goggle-eye" was almost identical with the plot of "Odisseus and Polyphemus" well known to the European reader from the Greek mythology and Homer's "Odissey". That was how von Diez called his extensive analysis of the Epos: "The Newly Discovered Oghuz Cyclope in Comparison with that of Homer's". Following his return from Türkiye, von Diez published this work of his in Dresden in 1815.

In his article von Diez treats "The Book of Dede Korkut" with great esteem. In his analysis he even considers Goggle-eye more ancient than Polyphemus as a character. He substantiates his idea with the fact that in Homer it is only from the moment of his meeting with Odisseus that Polyphemus is described and naturally it is incomplete. Contrary to that, in the Oghuz Epos the life of Goggle-eye or that of an Oghuz Cyclope, as he is called by von Diez, is described completely from his birth to his death.

It should be noted that this idea of von Diez, cannot actually resist criticism, since it is not correct. That is, it is possible to substantiate the incomplete description of Polyphemus in Homer. This is possible because Polyphemus had been known to the Greek audience well from the extensive Greek mythological context. Homer, simply, might have not needed that.

In general, this attitude of the German scholar towards the ancient Oghuz epos should be remembered with great gratitude. For the researchers of "The Book of Dede Korkut" the birth of the Epos has started since 1815 when Friedrich Heinrich von Diez' article was released.

2.3 The Customs and Traditions, Rituals in "The Book of Dede Korkut"

The customs and traditions peculiar to the ancient Oghuz community appear in a sufficient way in the Epos. The Epos is a text where ancient customs and traditions, rituals, ethical relations find their bright reflection in the background of the heroic descriptions. The relations between a father and a mother, a father and his child, a mother and her child, as well as between the lovers, the loyalty to the values that started their formation at the national level, the skill of resisting the external pressure, the love of religious values, the respect towards the preceding moral-religious beliefs manifest themselves vividly in the text of the Epos. All these are the issues to be studied even today.

As a moral chief—clairvoyant (shaman) of the tribe and further that of the community, the love and respect towards Dede Korkut himself, his word, advice, saz (a lute-like musical instrument), the obedience to the elders, adoring a mother even as much as Tengri, keeping one's promise are among the basic ethical features of this community. When dark clouds appear in the sky of the Motherland and during the enemy attacks, all feel obliged to arise unanimously to its defence. While an Oghuz warrior is in the captivity of the infidels, individual Oghuz commanders come to his rescue, gathering their troops. Children do not go against their fathers' statements. Unquestionably, the unity of the Oghuz is protected.

One can speak in particular about the hospitability existing in the Oghuz community and appreciated in the Epos. The notion of a guest is of very great importance in the Epos. We witness how much importance is attached to the notion "a guest" in the Oghuz community with this sentence in the Introductory part: *«Qonağı gəlməyən qara (böyük) evlər yıxılsa yey».* (*"Big houses that are not visited are better collapse"*). This idea is actually one of the ideological foundations of the Epos. No matter how big (here "qara" implies "big"), how rich (preceding from its being big) the house is, it is better the house not visited did not exist. Such kind of a house is better collapse.

When dividing women into four types in another place in the Introduction, the author, who condemns the type of women who do not respect the guests, thus, re-expresses how high the level of a guest stands. Displaying a respect to a guest is one of the brightest customs manifesting itself in the Epos.

The members of the community are mostly monotonous characters from the monumental, psychological aspect. Any of their actions can be accepted distinctly. They have a single member, they are the possessors of a single psychological nature. These characters who impress as a robot are either good or bad. Basat, Yaltajuk son of Yalanji, Aruz Koja and a horde of infidels' princes can be cited as examples.

Meanwhile, by and by some characters in the Epos start to acquire a binary psychological nature. It is already impossible to define their characteristic unambiguously. They attract the attention as a new type of characters for the text of the Epos. Beyrek, Lady Banu, and in some cases Salur Kazan are among such kind of characters. They have overt and covert layers. Overtly, they enjoy only positive features. Whereas, in the covert layer it is already possible to see some other psychological features of these characters. For instance, while being in prison, Beyrek, who is a bright representative of the new characters, establishes some contacts with the daughter of the infidels' prince to have a good pastime there. At this moment one can but not observe in Beyrek the formation of some psychological complexities. In this case Beyrek goes beyond certain frames of ethics and behaviour originating from the customs and traditions peculiar to the Oghuz heroes. This state should also be accepted as a proof that the text of the Epos was written in a binary literary esthetic "language". One of these "languages" is the language written under the dictate of mythological thinking, and the other one is the language in which writing traditions started steadily to manifest themsleves.

The ritual of "giving a name", which has already started dying away and thus losing its essence, is one of the most important rituals in the Epos. Sufficiently solid relicts of this ritual are manifesting themselves in the Epos. The boys until they reach the age of sixteen would not be given any names. All were expecting the young man to display some bravery till he reached sixteen. The names of young characters such as Bughaj, Beyrek and Basat are given after each of them displayed the related heroism. This ritual is carried out by Dede Korkut in person. It was Dede Korkut who gave the son of Dirse Khan the name "Bughaj" as the young boy killed a bull, to the son of Baybura Bey the name "Beyrek" for his courage (if expressed in a modern language, who had a liver) to rescue the merchants and their goods from the robbers, to the son of Aruz Koja the name "Basat" since the young man managed to handle a horse and ride it. Let us give an example. In the Epos, in order to give his son a name, Baybura Bey, one of the Oghuz nobles, *«Qalın Oğuz bəylərini çağırdı, qonaqladı. Dədəm Qorqud gəldi, oğlana ad qoydu, dedi, ünüm ünlə, sözüm dinlə, Baybura bəy, sən oğlunu Basam deyə oxşarsan. Bunun adı Bamsı Beyrək olsun. Adını mən verdim, yaşını Allah versin». ("Prince Bay Büre summoned the nobles of the teeming Oghuz and feasted them. He submitted his son's story to them and they approved. Then Dede Korkut came and gave the boy a name, saying, 'Hear my words, Prince Bay Büre, Your pet-name for your son is Bamsa; Let his name be Bamsi Beyrek. I have given his name; may God give him his years.'").*

This is how the ceremony of giving names takes place. With the process of giving a name, the admission of a young man to his community takes place. It is after Bughaj, Beyrek, Basat and other Oghuz youth have passed the ritual of being named that they could be accepted as a full warrior, member of the community with equal rights. In the Introduction of the Epos there is a moment when it is pointed out that Dede Korkut has given names not only to the youth, but also other items in the surroundings.

To carry out naming implied actually to rule the inner spirit of the ancient community. In this sense, it is not hard to imagine what was the role of Dede Korkut as a moral chief, and to what extent the power of the moral chief could reach.

One can point out as a moral ritual also the arrival of Dede Korkut to utter a final opinion and provide examplery advice when the main dramatic developments are over at the end of each chapter.

2.4 Language and Language "Disease" in "The Book of Dede Korkut"

Alongside with being the language of the sixteenth century Oghuz (ancient Azerbaijani), the language of the Epos maintains more ancient linguistic features as well. These features are associated with the failure of the language to generate sufficient strength yet.

Time and then among the researchers there have been disbutes whether the language of the Epos was prosaic or poetic. While, the language of the Epos can be considered neither as pure poetic nor prosaic language. Its language is the joint language of both poetry and prose where these two realities did not severe from each other yet. It is a kind of mythological centaur. This language is at the same time the language of both poetry and prose. Such a singularized language of "poetry-prose" is one more indicator of the antiquity of the Epos, we can say. During the further development this language is branched into the directions of poetry and prose. The language of "The Book of Dede Korkut" is the language of the period prior to this branching process. The most curious moment related to the language and style of the Epos is the principle of parallelism related to some design of the language units. We can encounter the most various forms of manifestation of parallelism in the epos "The Book of Dede Korkut". The general forms of parallelism such as syntactic parallelism, stepped parallelism, reverse parallelism, the parallelism of variants provide certain opportunities to imagine the literary structure of the Epos in a shape sufficiently polished. Syntactic parallelism is used to provide close relations of the sentences constituting the text with one another grammatically and by content. Parallelism is built on the structural and content identity or similarity of the sentences with one another. One can appraise this state at the sentence level as a transition from Chaos to Cosmos (from disorderliness into order). The striving for order and harmony is almost striving for harmony as a whole and at the language level of the Epos the most striking example of it is syntactic parallelism. For instance, let us consider such an example in the Epos. *«Qara-qara dağlardan xəbər aşdı. Qanlı-qanlı sulardan xəbər keçdi. Qanlı Oğuz ellərinə xəbər vardı. Uşun Qocanın ağ ban evi önündə şivən qopdu…». ("The news climbed the darkling mountains, the news crossed the blood-red rivers, the news came to the land of teeming Oghuz. Mourning began before Ushun Koja's white pavilion.").*

It is obvious that the grammatical structures of these sentences have been built in accordance with one another. Such kind of one-step syntactic parallelism is of great importance in the formation of rhyme and meter.

The type of syntactic parallelism which we can distinguish as "stepped" and naturally, its subtypes are the main form of narration of the Epos. We can call the subtypes as "one-step", "two-step", and "three-step". The example mentioned a little earlier belongs to the one-step parallelism. Here the sentences which maintain identity (or similarity) with one another through identical structures are used directly next to one another.

In two-step parallelism the identical (or similar) structures are alienated from each other at a distance of a sentence. The very sentences that occur inbetween agree with one another by the principle of parallelism. For instance, *«At təpdilər. Beyrəyin atı qızın atını keçdi. Ox atdılar. Beyrək qızın oxunu yardı». ("They spurred their horses and Beyrek's horse passed the girl's. They shot their bows and Beyrek's arrow split the girl's arrow.").*

During the three-step parallelism there is an alienation between the parallel (similar) sentences at a distance of two sentences. Again the inserted sentences inbetween enter parallel relations with one another in pairs. For instance, *«Qaytabanda qızıl dəvə bundan keçdi. Törümləri bozlayıb belə keçdi. Törümcüyüm aldırmışam bozlayayınmı? Qaraqoçda qazlıq at bundan keçdi. Quluncuğu kişnəyib belə keçdi. Quluncum aldırmışam kişnəyəyinmi?!» ("The camels in the herd have passed this way, Their baby camels with them, whimpering. I have lost my baby camel; shall*

I whimper? The Kazilik horses in the drove have passed this way, Their little foals with them whinnying. I have lost my little foal; shall I whinny?").

The reverse type of parallelism (chiasmus) is generated from the reverse use of the structure of the latest sentence from among the parallel sentences to the previous one. There is no example of such a complex stylistic point in the language of the Epos. The example is approximate. Given that the reverse parallelism belongs mainly to the written language, this implies that the language of the Epos is rather closer to the oral style, more precisely expressed, the lack of style. As for the rare and approximate examples occurring in the Epos, they can be appraised as the striving of the language of the Epos to Writing culture. We can give such an example, *«Nə gördü, sultanım?! Gördü, göy çayırın üzərinə bir qırmızı otaq tikilmiş». ("What do you think he saw? My Sultan, he saw a red tent pitched on the green grass.").*

Here the reverse order of the structure of the second sentence versus the first one can be proved (restored) only with the help of logical criteria. Therefore, we call such kind of examples of reverse parallelism in the Epos as the approximate examples. In our example the reverse order of the second sentence can be restored this way.

To fill the semantic gap of the element "what", which is the first word of the first sentence, the second sentence can be inserted into the first sentence as a whole. In other words, we can state that in this model the visibles of the second sentence ("*a red tent pitched on the green grass.")* is the correlate of the pronoun "what" in the first sentence.

We can restore the address "my Sultan" in any place of the sentence. In our case let us restore it at the beginning of the second sentence. After the restoration the example of the reverse parallelism will have the following view: *«Nə gördü, sultanım?! Sultanım, gördü, göy çayırın üzərinə bir qırmızı otaq tikilmiş». ("What did he see, my Sultan?! My Sultan, he saw a red pavilion was built on the green grass.")* Such kind of approximate chiasms suggest that the language of the Epos is not completely ready yet to describe the subtleties of Writing culture.

One can also speak of the parallelism of the variants in the Epos. The parallelism of variants manifest itself more often at the level of content. This implies the similarity or identity of one or another action of characters, the similarity of some plotlines in different chapters. For instance, the general, invariant content associated with Salur Kazan's captivity is "scattered" to different chapters as variants characterized by identical or similar features. The situations such as hunting, wedding, description related to both the captivity of the hero and his rescue, the relatives' recognition of each other before the battle, struggle, competition, assault are presented as similar variants in different chapters.

It should be mentioned that the parallelism of variants can already surpass the boundaries of the text of the Epos. The comparison with the ancient Greek mythological texts shows that one can encounter similar or identical variants displaying themselves at the content level of the eposes created in different areas. It doesn't matter which of them had been created earlier and which at a later period. That is, the idea that the variants in both the areas have originated and spread from a single more ancient invariant source can be considered satisfactory and productive for both the variants.

One of the utterly significant issues manifesting itself in this ancient Oguz language is an issue related to "language disease". This point supported by the mythological vision is typical for the environment surrounding an ancient man. When language finds it difficult to name a new notion, it is obliged to name it with a sign already at its disposal (within its arsenal). And we shouldn't forget that the original meaning of this sign is still alive in mythological vision. For instance, in this connection let us consider the expression "«qarşı yatan Qara dağ» ("there lying black mountain yonder"). The first thing striking the mind is the verb "yatmaq" (in English "lying") which seems to have been incorrectly used. Its primary meaning is visible from the substratum of the sign. The ancient man fixes in his brain not only the mere vision of the mountain, but its being asleep as well. And he believes in it. He is convinced that the mountain asleep will one day wake up. Mythological vision interfers his consciousness this way.

However, it should be taken into account that when the text of the Epos was created, this sentence needed a verb like "being situated" (in order to get the structure "the mountain situated ahead"). Whereas, as it did not exist on the language plane yet, the verb "asleep" comes to help and replaces it. Since, in the cave every night the author of the Epos used to see a co-caveman fallen asleep and remaining motionless. This fellow caveman would wake up in the morning. Therefore, the caveman was convinced that mountain would also wake up. It is due to the disease of language (actually, its poverty in antiquity) that the verb "asleep/lying" comes forth and replaces an element not existing in language yet. Naturally, it brings along an important component of its semantics as well, does not allow it to get forgotten. This is the semantics "to sleep". Mythological vision originates from here and generates power.

Myth and Writing in "The Book of Dede Korkut"

In the Epos we sometimes see the mythological visions and the manifestation of writing culture next to each other. The prehistoric stage of the ancient, primitive man's understanding of the world is the very mythological stage. Searching for the features pertinent to himself and "finding" them form this man's morals and worldoutlook. In the language of the Epos the signs of this ancient period remain frozen as relicts. If we encounter the expression *«Qarşı yatan Qara dağ»* (*"there lying black mountain yonder"*). In the language of the Epos, it implies that when this expression was being formed, they were indeed convinced that the Black (Great) Mountain lying ahead was asleep and would wake up one day. Ascribing a feature pertinent to himself, i.e. sleeping to a mountain which belongs to Nature is the brightest expression of the antiquity of the language of the Epos. Mythological world is an outcome of the mind of a man who created the Epos. In this sense, the expression *«de, baxalım, nə deyirsən» ("say, let us see, what you declaim")* can be appraised as a manifestation of accepting the utterance, i.e. the word as a value possessing a real, physical volume. The ancient man who is a possessor of a mythological mind is convinced that the mountain lying yonder will wake up some time till he reaches the threshold of literary thinking. He is convinced that the soul that comes out will return. He is convinced that the dying man will resurrect, and so on and so forth.

As a human being develops, as the process of his comprehension of the surroundings, generation of parallel worlds (as plots) in his brain discovers the features similar to the reality, the period of artisticism begins. Writing culture raises its head furtively from under the mythological visions. At that time even mythological plots can be exposed to literary refining. "The adornments" period of writing culture starts.

The period of writing starts with the appearance of writing itself and gets established. With its formation writing happens to bring various new codes to human brain. The word, which is the absolute ruler of Myth starts to grow invisible within the sentence. The general content, weight of the sentence which is the main foundation of Writing culture, dissolves the Word within its womb, turns the Word into a word. Thus, it is the sentence that takes the weight of the text onto its shoulders. The refinement process starts operating at different language levels. A tendency towards stylistic variety gets consolidated. Metaphors, exaggerations, adornments already turn into constant attributes of the text. The reign of the particle "kimi" starts. Writing culture opens new, extensive opportunities for such texts as "The Book of Dede Korkut". The prenatal life of the Great Fizuli's magnificent line *("Don't get deceived that the poet's word is certainly a lie")* starts.

In the text of "The Book of Dede Korkut" we see the elements of mythological imagination and the elements of writing culture side by side. However, one should certainly not forget either that the last formation stage of the Epos took place under the impact of the period of the renaissance of Writing culture. Besides this, it should be noted that mythological elements were not easily wiped out of the text either. Even sometimes finding convenient shelter for themselves in our language, these elements have survived till our days. For instance, the expression *«de, görüm, nə deyirsən?!» ("say and let me see what you declaim")* found in the Epos is the form of the expression *«de, baxaq görək, nə deyirsən?!» "say, let us have a look and see what you declaim?!"* surviving even today in our modern language. The word, which has turned already into a cultural value this way, appears as a real, personified creature, as a visual creature as it first used to be understood, in other words, as a creature outcoming from mythological imagination.

2.5 Discipline, Order and Systematicity in "The Book of Dede Korkut"

We sometimes get any (maybe important) information delivered from the Epos not from its overt layer. The author seems to hide it from us. The information is not presented in particular where it is supposed to be. The author transforms it to another unexpected layer. And it is sometimes transformed to another chapter. This state, that is, 'the concealment" of the information, its discovery on another narrative plane can also be appraised as the manifestation of fuzzy logic. This also provides more solid relations among the chapters. The fate of Beyrek, one of the characters of the Epos, can be a bright example of it.

Beyrek is the only warrior who is condemned to death by the author and whose death is described in the Epos. The reason for his death is related to his certain action in another chapter. In this other chapter Beyrek has been in captivity of the infidels for 16 years. He escapes from this captivity with the help of a daughter of the infidels' prince and when fleeing he promises the girl to return and marry her. He swears an awful oath: *«Qılıncıma döğranım. Oxuma sancılım. Əgər Oğuza sağ-salamat gedib çatarsam, geri gəlib səni halallığa almazsam...» ("May I be sliced on my own sword, may I be spitted on my arrow. ... if I reach the Oghuz land safely and do not come back and marry you.").*

Beginning from this point the plotline corresponds to the developments that occur after Odisseus returns to the Island of Ithaka. Like Odisseus, Beyrek also returns safe and sound to his native land. Nobody recognizes him when he returns, even his sisters. On the day when his fiancée is about to marry another man, Beyrek visits the wedding and wants to distinguish his friends from his foes by remaining anonymous. In this very anonymous way he visits the wedding gathering. Odisseus also finds himself in such a situation on the Island of Ithaka. He also gets to the wedding gathering of his wife Penelope. The youth who claim to possess Penelope's hand and heart, actually her throne do not recognize Odisseus.

During the wedding Beyrek has to undergo some local tests and finally, he introduces himself to those who don't know him (absolutely like Odisseus), and is reunited with his fiancée Lady Chichek. Penelope does not recognize Odisseus. Likewise, Lady Chichek first does not recognize Beyrek either. Like Odisseus, Beyrek also reveals to his fiancee some secrets of their engagement times and thus, the issue is solved. Odisseus' secret is associated with the bed in the bedroom. Nobody knows this secret except him and Penelope. It was impossible to remove the bed, since it had been made on the stem of an oak tree and its feet stretched to the depth of the soil. Whereas, Beyrek reminds Lady Chichek about his victory during their competition and "kissing her thrice and biting once, and their loving each other". These remindings are enough for the lovers to know each other and there is no obstacle left and the hour of lovers' reunion arrives.

In the Oghuz mythological space Beyrek did not keep his promise made by him a while ago that he would return to marry the infidel's daughter if he reached the Oghuz land safe and sound. He swore an oath and in accordance with the mythological principles Word (the Word that wrote the Epos!) punishes him. That is, in another chapter owing to his loyalty to the prince of princes Salur Kazan, Beyrek is killed by the princes of the Outer Oghuz rebelling against the Inner Oghuz. Why is the "hidden" information related to the murder of Beyrek set in another narrative space of the text of the Epos in complete accordance with the principles of fuzzy logic, and arises from there?

Discipline and order in the text of the Epos regulate, above all, the hierarchy in the content. At Bayandur Khan's gathering this hierarchy is described as follows. "The right princes" sitting to the right and "the left princes" sitting to the left, the favourites standing outside differ from one another.

The selection of opponents prior to the battle is extremely important. That is, the fighting of Sher Shamsaddin, one of the Oghuz heroes, "without consent", going against Bayandur Khan's opponent, is not accepted with respect by the Khan at all.

During the battle each Oghuz hero fights against his own opponent as agreed beforehand. This is one of the principles of the fighting.

The chapters have deep and close interrelations. The reason presented in one chapter appears as an outcome in another. This state also testifies to the solidity of the interchapter relations. Consequently, the idea that we encounter not twelve separate chapters, but a single whole text does not leave the reader.

2.6 Homer and Dede Korkut

The magnificent monuments created by these two great poets describe the period when ancient mythological visions are close to each other even in the most distinct areas. Therefore, these works cover the description of sometimes sufficiently similar and identical points.

In the light of the above stated, it would be interesting to compare the battle scenes described in "Illiad" by Homer and in "The Book of Dede Korkut" by Dede Korkut. This comparison would allow us to present how the poles of fuzziness and precision behave on the stated plane (in the description of battle scenes) in these immortal texts.

During the description of the battles we observe who overcomes whom in both the eposes.

One of these poets displays extensive details in the description of the battle scenes, providing information about the very family members of the dying and killing Greeks and Trojan heroes—this is Homer. In the battle descriptions Homer is more comprehensive. For instance, while an entire chapter (song) is dedicated to the brave Diomedes' acts of heroism in "Illiad", it is sufficient for the Oghuz author to just mention the victory of an Oghuz hero over the enemy in front of him during the battle in "The Book of Dede Korkut". He declares it in a sentence. He doesn't feel any need for details.

With patience and enthusiasm Homer describes how and in which sequence the warrior wears his combat gear prior to the fighting. Paris wears his uniform prior to his fighting with Menelaus in such a sequence. He first puts on his bright knee wears. Then he fastens them tightly to his legs with ties of silver cords. Further, he wears his apron. He puts on his sword. He takes hold of his huge shield. He puts on his thick headgear on his head. He takes his lance in his hand. Following this concrete and lengthy description, even Homer feels fed up and introduces in a line in a "Dede Korkut" style the adornment of Menelaus, who is Paris' opponent in the battle scene and says, "Menelaus was also armed and clad like him".

In "The Book of Dede Korkut" the accurate confrontations are defined prior to the fighting between the challengers. If the warrior wins his opponent, that is his challenger, the description in the text of the Epos is over. Whereas, in the indirect description of fighting Dede Korkut seems to be in a hurry, allows some reductions with the Occam razor, simply delivers the message on "who" out of the opponents defined prior to the fighting "killed whom". The very state like this suggests that the fighting itself as a whole takes place not in a chaotic way, that is "whoever found whomever wherever it comes from", as is the case of Homer, but in a more orderly (planned) way. Such 'an orderly" description in "The Book of Dede Korkut" testifies to the principle of making a concrete choice in a fuzzy way.

Whereas, in Homer the "chaotic" description manifests itself as an effort to embrace the entire battle scene. Homer's comprehensiveness and Dede Korkut's conciseness oppose each other and here we witness once again that the text of Dede Korkut is an outcome of fuzzy thinking.

Thus, we see that in the fighting scenes Dede Korkut acts as a whole by the principle of fuzzy logic and provides approximate descriptions.

Whereas, Homer, contrary to this, tries to remove fuzziness, and reflect the reality as a whole.

Certainly, if that was the issue, there is no need to get surprized why the great Aristotle, who was the product of the antique world, could not view the world through fuzzy thinking. The literary model familiar to him and which he grew up with was the model of clear thinking. Aristotle could not have accepted the existence of a third between two opposing poles.

One can also point out another distinguishing feature. Homer's inclination to details and Dede Korkut's effort of precision (actually, his inclination to fuzziness) appear in the heroes' presentations as well.

Homer counts almost all the ships of the brave Akheys who swam along the Trojan coast, mentions the main commanders'-in-chief. Homer has a concrete answer to the question, "How many ships did each commander-in-chief take with him? Again here one cannot but admire Homer's preciseness. The number of the ships sometimes can be hundred, sometimes fifty, and sometimes forty, twelve, seven, thirty two, even three times more. Homer doesn't use such hedges as "in general", "many", "considerable number". Such preciseness of Homer can also be assessed as an effort to shun uncertainty.

Homer's inclination towards the details manifests itself in the poem "Odissey" as well. We witness it in Odisseus' meeting with ghosts in the land of the dead. Odisseus who meets with the spirit of the dead talks to them one by one and meanwhile Homer does not forget either to provide sufficient extensive information about the extended family, forebearers of each of them. The talks with Achilles and Agamemnon's spirits can be emphasized. The information Oddiseus gave him on Achilles' son, the information given by Agamemnon to Odisseus on the murder of Agamemnon on his way back to his home by his wife and her lover Egist are striking. Such an inclination to the details, once again proves that Homer "is working" in the regime of crisp logic.

Whereas, Dede Korkut surfices himself to just listing some (!!) of the Oghuz heroes who come to each other's rescue when in trouble. It is true that Dede Korkut does not forget either to mention this hero's only one characteristic feature well known in the land of the Oghuz. For instance, it is said about Sher Shamsaddin that this Oguz warrior has attacked and killed Bayandur Khan's enemy without asking

Bayandur Khan's permission. After providing such laconic information about each noble, Dede Korkut creates a decent set system with the hedge *«saymaqla saya gəlməz» ("of counting ... there could be no end").* He does not feel any need to mention all the Oghuz nobles by their names. Again we cannot but feel here the demonstration of Dede Korkut's inclination towards fuzziness.

World-renowned Argentine writer J. L. Borges has such a wonderful story: "Funes—the Memorius". This story deals with an extraordinary and magnificent skill of Funes, a young man. Funes can record to his memory everything taking place around him. He can repeat all that happens within 24 h one by one within another twenty four hours. There is no notion of hedge for Funes. He can present the surrounding world as a whole. Funes is as far away as "a black gap" from fuzzy logic. Funes is the follower of Homer's logic. Whereas, Dede Korkut couldn't have accepted the phenomenon of Funes.

2.7 The Study of "The Book of Dede Korkut"

It should also be noted that following von Diez' introduction of "The Book of Dede Korkut" to the scholarly world (1815), beginning from the early nineteenth century its study process has started in scientific centres in Türkiye, Germany, Russia and other countries. In 1950 another edition of the Epos (six stories) was found in Vatican. Following that, the Epos was translated into Italian by V. Rossi, an Italian scholar in Eastern Studies.

Thus, beginning from the nineteenth century Turkish, Russian, German, Italian scholars have started to study the epos. Among them one can mention the Turkish scholars Kilisli Muallim Rifat, Fuat Köprülü, Muharrem Ergin, Orhan Şaiq Gögyay, Ahmed Bican Ercilasun, Osman Fikri Sertqaya, Fikret Türkmen, Ali Duymaz, Ali Erol and M. Ekici, who have enriched Korkut studies with their valuable researches.

Among the Russian scholars V. V. Bartold, V. V. Zhyrmunsky, A. Kononov, Kh. Koroghlu, L. Stebleva can be distinguished.

The Azerbaijani scholars such as Amin Abid, Hamid Arasli, Mammad Arif, Mammadhuseyn Tahmasib, Ali Sultanli, Əbdulezel Demirchizadeh, Shamil Jamshidov, Farhad Zeynalov, Samet Alizadeh, Tofig Hajiyev, İsa Habibbeyli, Rahman Badalov, Vagif Ibrahimoghlu, Aydin Talibzadeh, Niyazi Mehdi, Arif Ajaloghlu, Kamil Veli can be distinguished for their contributions to the study of "Dede Korkut" more or less.

The Epos is being studied fundamentally from the aspects of language, literature, history, ethnography, cultural studies.

One of the latest interesting studies is "The Book of Dede Korkut" and Fuzzy Logic" co-authored by Professor Kamal Abdulla and Professor Rafik Aliev and now in the hands of the reader. Three editions of this study have been released in Azerbaijan. In Türkiye it has been published in Turkish by «Ötüken» Publishing House, and its English variant is under publication by the international publishing house ELSEVIER.

"The Book of Dede Korkut" as an Object of Repression

In 1950 Hamid Arasli and Mammadhuseyn Tahmasib published V. V. Bartold's translation in Baku following which a widespread repression of the Epos started. In this period a complete tabu was imposed on the study of the Epos. Its study in the Turkic republics, especially in Azerbaijan was banned. Its publication, dissemination, expression of opinions about the nation's history, psychology, worldoutlook, language, ethnographic features on its basis, briefly any research related to this Epos were forbidden. An identical state took place in Turkmenia, another Soviet Republic. There the Soviet regime "worked" more severely. There the researchers of "The Book of Dede Korkut" were thrown into prison and in some cases even condemned to death.

The leader of the Azerbaijani Bolsheviks Baghirov forced a number of Azerbaijani scholars to a "crusade" against the Epos with his political speeches in party conferences. He labeled this Epos as a harmful work damaging the friendship among peoples, contrary to customs and traditions, insinuating the alien lifestyle, glorifying the senior strata of the Oghuz tribes who invaded Transcaucasia, incorporating the bourgeois nationalism. Following these speeches, a real fighting started against the scholars studying and advocating the Epos. As we have just pointed out, the main blow came from the political direction.

The repression against the Epos can be divided into two stages. The first stage attracts the attention as a "voiceless" stage, and the second one as a "voiced" stage.

In the first "voiceless" stage the Epos, as we can state, accompanies the period of historical formation of the nation from the literary aspect. However, the issue is that until von Diez' publication neither in Europe, nor in the Muslim world the Epos had not been known.

We would not be wrong, if we say that in the period prior to von Diez' introduction this Epos had been exposed to hidden (and voiceless) repression of Muslim ideology. Its penetration to existing madrasas, scientific institutions, and wide masses actually was forbidden with various obstacles. According to the researchers, even the traces of Muslim culture patched later on to the text of the Epos with colourful yarns (Ali Sultanli) failed to prevent this repression forced against it. We can note that the reason for that was the lack of chance to remove the fragrance of paganism from the Epos. There were actually no Islamic values in the deep roots of the text. That is, it is impossible to find not only Korkut's real trace, even any signs of his spirit in the works of many, maybe, majority of, or maybe without any exceptions, all the classics of Azerbaijani literature—Nizami, Khagani, Nasimi, Khatai, Fuzuli, Vagif and others. These literary figures were unexceptionally educated, wise men of their periods. It turns out that there is not any information on the Epos even in the remotest corner of madrasas that were considered the arsenal of knowledge. The Epos has not simply been concealed. The Epos has not simply been acknowledged. As the culture of "Dede Korkut" (it is sometimes referred to as the culture of Bayandur) had not been adjusted to Islam, the culture of Bayandur pertinent to the spirit of Korkut remained quietly asleep under "the captivity" of the Islamic spirit reigning similarly in the works of these classics.

The second period of the repression can be distinguished as the "voiced" (maybe rather "with shouts") period. We call it "Bolshevik repression" as well. This repression, which remained secret until 1950, began "to receive" its sacrifices with the release of Bartold's work in Azerbaijan. Moreover, they started the persecution of these sacrifices not only in Azerbaijan, but also in Turkmenia. The Party leaders organized ideological attacks against the Epos there as well. Karriyev and Kosayev, the scholars studying Korkut, were arrested and exposed to tortures, Kosayev was even condemned to death.

As the main reasons of the second period of repression, its incorporation of the aggressive bourgeois nationalism alien to Soviet ideology, presentation of a lifestyle, customs and traditions allegedly alien to the Azerbaijanis, glorification of the Oghuz' upper strata, and its instigation of discords among the Transcaucasian nations were underlined.

There was also an invisible, secret feature related to repression. It wasn't in vain that the epos "The Book of Dede Korkut" infuriated the Bolsheviks. It shouldn't be forgotten that the essence of the Epos towers over the ethnic-ideological unity of the Turkic nations. This unity brings forth the identity of the lifestyle, worldoutlook, customs and traditions, especially, language identity (similarity). Such unity of the Azerbaijani people with the Turkic peoples living in the Soviet Republics, especially the Turks living in Türkiye was a perspective fundamentally contradicting the Soviet ideology. It was a perspective leading to the great Turkic unity which the Soviet ideology wouldn't allow. It is here that one should hunt for the genuine reason for the voiced, noisy repression, we surmise.

It is due to this very reason that an example of folklore which encorporates many features of the people's ethnographic life, manifesting itself as a material of antiquity at a wonderful literary level, describing the points full of heroism with a supreme literary taste, playing a part of substantial source to study ancient language elements became the target of the fury of the Soviet ideological system.

Although the language of "The Book of Dede Korkut" is closer to some of the Turkic peoples and remoter from the others, it was the language of a text written in the same degree of understanding. While the appearance of such kind of language identity (or similarity) could be considered relatively accepted for the peoples living within the Soviet empire, such a state could absolutely be unacceptable from the point of view of Soviet ideology if compared with the Turks living in Türkiye beyond the boundaries of Soviet Empire. Since, the Epos was a work written on a common language level of the geneologically closest languages—Azerbaijani, Turkish, Turkmen, and Gagauz. Such kind of proximity, was clearly not compatible with the interest and views of Soviet ideology at all.

Thus, the main reason for the exposure of the Epos to represession, undoubtedly, was the fear of revelation of the identity on a language plane, in customs and traditions, ethical-moral principles, thus following these identities, naturally the initiative of removal of the danger in advance for the national identity to find ways to ideas, credos [1].

2.8 Democracy in "The Book of Dede Korkut"

One should distinguish between overt democratic elements and covert democratic embryos in the text of the Epos. The overt elements are those visible on the surface of the text. For instance, while fighting against Goggle-eye every family was supposed to give him a child to eat. This division was ascribed to the entire Oghuz community. The Epos says that those who had four sons, gave away one and three remained. Those who had three sons gave away one and two remained. Those who had two sons gave away one and another one remained. The turn could rotate and again reach the family with no child left. Though at the expence of giving away the captives, the family was supposed to give this debt. No family, no matter who they were, what positions they were holding within the community, could go beyond this division.

Another example appears during the conversation with the nobles regarding Salur Kazan's frequent address to them as "Nobles!" This address turns into a kind of key word of the intention of Salur Kazan—The commander-in-chief of the Oghuz people in regard with his getting the consultation of the nobles. It is after this very address "Nobles!" that the Oghuz nobles start to speak and make their suggestions one after another, which indicates the unanimous support of the adopted decision.

Or such a simple example, we think, suggests the presence of democratic tradition. On the eve of the wedding a silk shirt is being sent as a gift to Beyrek by his fiancée Lady Chichek. Beyrek's forty fellow-warriors get disappointed at the sight of it, since their shirts are not so beautiful. Having sensed this, Beyrek solves the puzzle as follows, "Today I wore it. Tomorrow each of you will wear it by turns. Then we'll give it to a derwish."

This is an impeccable democratic approach for the described period. All these are the external signs of the manifestation of democracy. Whereas, the hidden embryos are in the hidden form in the depths of relations of language and thinking.

The hidden democratic embryos are mostly discovered by the examples of the elements of fuzzy logic scattered to the text of the Epos. As an example, one can state that in the text of "The Book of Dede Korkut" (especially in a very scattered state) a third, even a forth and sometimes a fifth can be inserted inbetween the two poles filled with an opposing statement, unlike Aristotle's logic. A choice is expanded for the recipient, thus a genuine democratic vision manifests itself. The presence of such moments in the mindware of the ancient Oghuz testifies to their attempt to see the objective reality in its adequate variety.

The presence of democratic embryos in the community described in the Epos is also confirmed by such a fact. When the democratic "rules" typical for the described ancient period are violated, dissension breaks out. The Oghuz warriors do not forgive each other for such a state. The real reason for the fight that takes place between the Outer Oghuz and the Inner Oghuz is the violation of the ancient democratic principle. This principle is the following. Once a year Salur Kazan, the commander-in-chief of the Great Oghuz, lets the princes of the Inner Oghuz and Outer Oghuz, which are the parts of the Great Oghuz, pillage his house. That time he would take his wife Lady Burla the Tall by the hand and leave his home. The princes of the Inner Oghuz and

Outer Oghuz would enter Kazan's house and take from the house whatever they liked. This was, clearly, the primitive manifestation of ancient democratic relationship in the Epos.

Once Kazan violates this "democratic" custom and while he is having his house pillaged, he doesn't invite the Outer Oghuz nobles to this ceremony. The reason is not known. Consequently, the Outer Oghuz princes are annoyed from Kazan Khan and naturally, they turn hostile to the Inner Oghuz and rebel against the main power. A fighting breaks out between the Inner Oghuz and the Outer Oghuz. In this battle Kazan defeates the Outer Oghuz princes and their chieftain, horse-mouthed Aruz Koja. Although the protest against the violation of the democratic custom is subsided, the occurrence of this protest by itself confirms indirectly the existence of that democratic principle.

Thus, the above-stated results in a conclusion which causes a surprise at first glance that the Outer Oghuz princes' efforts to start a dispute actually turned out to be their attempts to restore democracy. The Outer Oghuz princes who even allowed to kill Beyrek for not taking their side, if we consider the situation comprehensively, are actually carrying out the mission of mythological visions to punish Beyrek on the one hand, and they are shedding blood to restore the democratic custom on the other, we would say. Such an approach allows the reader to appraise from a different perspective the Outer Oghuz uprising which has so far been presented as an effort to violate the Oghuz unity by the Outer Oghuz princes. We can accept this uprising as the first struggle of the ancient Oghuz for democracy. Though in a chaotic form, Aruz Koja and the Outer Oghuz princes are protesting against the removal of a democratic custom. And in this regard, Aruz Koja attracts the attention as an Oghuz hero who rises to wage a life-and-death battle against Salur Kazan who has violated the democratic custom.

The epos "The Book of Dede Korkut" holds a significant position in the world system of mythological epos creation. The similarities of its many plotlines with the poems of Homer, "Gilgamish", the epos "Song of the Nibelungens", which belongs to a later period, allow us to state that historically the Oghuz epos is a text which started its formation in a rather ancient mythological period, passed through different editorial efforts time and then, modellized in literary-artistic aspect a transition from one into another in the pair "Nature-Culture" even in its own example, adequately described the processes ongoing in the consciousness. In this text crisp logic says its own important piece, likewise the manifestations of fuzzy logic make their presence known too. Fuzzy logic manifests itself in its different facets (foundations) in this Epos. We clearly see that remaining loyal to the feelings of tolerance and softness present in their mind, our ancient ancestors established their lives accordingly, tried to view the world, the surrounding environment from a democratic angle. Various manifestation forms of fuzzy logic in the Epos bring forth our ancestors' attempt to see the objective reality more adequately. This attempt, has definitely, had an impact on his worldoutlook, psychology, consciousness. One should also search here for the roots of the present-day multicultural mood that has turned into the characteristic of our people.

Beginning from the mid 1950s the ice of the repression regarding the study of "The Book of Dede Korkut" started to break and thaw out, and an extensive and frontal study of the epos "The Book of Dede Korkut" was launched in all the Turkic republics, especially in Azerbaijan. Today its language and style, historicism, philosophy, ethnographic features are being studied with great enthusiasm and interest. And finally, today with this book, which is being presented to the reader, the study of the text of "Dede Korkut" has been initiated in terms of the principles of fuzzy logic, we can say.

This way "The Book of Dede Korkut" and "fuzzy logic" attract the attention for the first time like two mighty rivers flowing next to each other. More precisely, these rivers do not flow next to each other, they are continuing to flow next to each other within one bed, yet each of them in its own colour. This can be considered the initial mergence. A precautious and fearful geniune mergence is to come a bit later. Today these two values have already started " to thaw out" within each other on a single plane of scientific view.

There is a site called Sugovushan (the place where waters merge) in the region of Sabirabad, one of the picturesque nooks of Azerbaijan. As it is obvious from its name in Azerbaijani, Sugovushan is a place where two huge Caucasian rivers merge and set forth together in the direction of the Caspian. The grey colour of the water of the Kurr is a normal, ordinary river colour. Whereas, the colour of the Araz is different. It comes from the mountains and maintains its somakh-like colour, the colour of the mountains through which it is flowing. After these two rivers merge, they continue flowing next to each other like this for a while. Meanwhile, it is possible to distinguish at a glance the Araz in the somakh-like colour from the grey-coloured Kurr. However, after a while, after they make approximately a kilometer, these giant rivers thaw out within each other, each losing its own colour, it is at this moment that the spirit of a single river and the genuine mergence in the colours take place. When they join the Caspian, they do it as a single river.

Likewise, one can point out the text of "Dede Korkut" and the principles of fuzzy thinking. It is both curious and useful to study them within each other, compared with each other, to apply the principles of fuzzy mindware to the Epos.

They are like two giant rivers. While applying one into another (fuzzy logic and mindware into the text of the Epos), these two magnificent rivers initially will continue flowing within their own beds each, but already along each other. And after a while, their genuine union, thawing out within each other take place. Now we even can be surprised. How could it be that we happened not to have noticed earlier this leading mindware in the text of the Epos, the mishmash of the relations encompassing the principles of fuzziness?!

By the end of this part, what remains to us is to note that the founder of fuzzy logic Lotfi Zadeh would wish to have his theory to initially be applied to the field of social sciences. From this point of view, the book presented to the attention of the reader is to a certain extent the realization of Lotfi Zadeh's dream, we would say. Such application to the example of an ancient language and thinking like "The Book of Dede Korkut" will lead to the searches of fuzzy logic in other literary peaces as well, we think.

The authors hope that the principles of fuzzy logic will manifest themselves in other pieces of our folklore as well. Such an application would rather enlarge the boundaries of our literary mind and we are convinced that it would bring useful perspectives also to the field of fuzzy logic. In this book we tried to define as much as possible the theoretical and methodological ways of those future applications.

Reference

1. Abdulla, K.: The Semantic Twinkling in "The Book of Dede Korkut", p. 164. Mytholinguistics. Baku (2022)

Chapter 3
The Elements of Fuzzy Logic in "The Book of Dede Korkut"

If we place "The Book of Dede Korkut" on one plate of the scales and the entire Turkic literature on the other, they will weigh in favour of "The Book of Dede Korkut".
Prof. Dr. Fuad Köprülü
Fuzzy logic is not fuzzy, in large measure, Fuzzy logic is precise.
Prof. Dr. Lotfi Zadeh

3.1 From "Either …Or…" to "Neither…Nor…"

The epos of heroism titled "The Book of Dede Korkut" is a literary text created by our ancient ancestor since the period of caves (pre-literary history) [1, 2] till the fifteenth century (until the time when it was recorded) step by step, word by word, sentence by sentence, episode by episode, point by point. The epos is, actually, of encyclopedic nature, and its overt and covert layers are displaying our nation's customs and traditions, culture, language and even their development (changes). Also, the worldview of an ancient man (our ancestor), his/her attitude towards the surrounding world, his/her vision of himself/herself and his/her own society are reflected in the structure of the language and content of the epos. The mindware of the narrators of the epos, the synergetic perspectives opened by this mindware before our ancient ancestor present an interesting and promising study object for the theoreticians of logic and information. And it is not a coincidence that the theory of fuzzy logic, which was introduced back in the mid-1960s and has accumulated more and more proponents over the years, is inclined towards social sciences according to the desire of its founder, professor L. Zadeh. It is in full harmony, first and foremost, with the text of "The Book of Dede Korkut".

Who was L. Zadeh and what kind of innovation did he introduce to the realm of science?

In scientific literature the well-known Western scientists divide the great scholars relatively into two categories [3]. Romantic scholars and classic scholars. Being a generator of ideas, the former offer a number of original theories all through

K. Abdulla and R. A. Aliev, *The Book of Dede Korkut and Fuzzy Logic*, Studies in Fuzziness and Soft Computing 438, https://doi.org/10.1007/978-3-032-05906-2_3

their lives. Thus, they kind of lay foundations for the building of science. Further, thousands of scientists in the world build up this construction.

The latter are the architects of this construction and by their efforts this building of science is constructed very solidly and fundamentally. We can state without hesitation that L. Zadeh, who is the author of five fundamental theories, is a person belonging to the group of romantic scholars. We would like to point out some features characterizing him as a genius.

Independence. We see it in his non-standard nature, his out-of-the box thinking, his being free from hesitation when confronted with the influence of others, and his ability not to yield to old ideas deep-seated in minds.

Flexibility. His freedom of not yielding to the existing rules and templates, as well as his ability to accept the old experience and draw a necessary outcome testify to his flexible scientific thinking.

Utmost tolerance. This is one of his best qualities. L. Zadeh demonstrates an attitude of tolerance towards complex, uncertain, and unyielding problems and cases. He enjoys a very strong ability to see the hidden regularities in the obvious chaos. Likewise, L. Zadeh also has a strong urge for originality and innovation. A highly developed esthetic sensation, sensitive attitude towards beauty and harmony are among the qualities characterizing his morality. It is not a coincidence that the fuzzy logic which maintains tolerance in its gist is the outcome of the intellect of this genius.

It should be noted that complete silence had not been observed regarding the accurate logical principles reigning in the minds prior to L. Zadeh either. Now and then incoherent "protests" unable to rise to the level of a theory were leaking through the borders of logic either from philosophy or physics or the theory of sets. Yet, it was not sufficient. The objective world needed a theory capable of providing its adequate description. And this theory was introduced by Zadeh [4].

With its various manifestations fuzzy mindware has merged into a scientific line called 'fuzzy logic" and actually been created as an extension of Aristotle's formal logic which had governed our mind until then. This logic is a mindware far beyond categoricalness, much closer to a human being, his/her principles of understanding and acceptance of the world. Those, who have discovered it, have done so "not in vain". Human beings have been a kind of sensitive to fuzzy mindware since the time when their mind was shaped. They perceived that the Procrustean bed of clear (formal) logic wouldn't allow to see the subtleties existing between the poles of good and evil, hot and cold, smart and crazy, experienced and dilettante, white and black (one can cite infinite binary sets) in our ancestor's mind. That is, proceeding from his own life experience, Aristotle stated utterly not accepting any protest: this man is either dead, or alive. A third was excluded. The choice consisted of the preference of only one of these two poles that accepted no protest. A human being could be either dead or alive. There was no other alternative. In fact, between the poles "dead" and "alive", a dead gap was slipping away towards infinity. The diversity of the objective world did not interest Aristotle and his followers. Only Plato made an initiative to point out a third position in this gap. However, his idea was not attractive. In the recent past, in the 1920s, the Polish scholar Jan Lukasiewicz suggested an idea of a trinary logic between these opposite poles. Further, an addition "maybe" introduced to the bipolar logical chain by the Hungarian-American scholar J. von Neumann was also put forward [5]. We could rather say that although not being out of the focus of attention, all these, i.e. the additions made after Aristotle failed to find their serious and consistent continuation. Again a big gap was being felt between the poles.

Finally, in 1965 the Azerbaijani-American scientist L. Zadeh developed and advanced a theory of continuous-valued logic [4]. L. Zadeh suggested an idea of the existence of an infinite number of shades (variants) between the poles *"true"* and *"false"*. These shades (variants) were *partially true, partially false* and their infinite modifications. This way, almost like in Zeno's famous paradox, the fleet-footed Archilles can never take over the tortoise if the latter is a meter ahead. Thus, the shades between the two poles won't end. They remain visible and invisible with infinite modifications. The gaps can be filled till quantum. The direct co-relation between fuzzy logic and quantum mechanics allows to state that the fuzzy logic, which has acquired the ideas of the Copenhagen school regarding the division of an atom, negates to some extent Aristotle's logic from this perspective as well.

As mentioned a bit earlier, while founding the theory of formal logic Aristotle and his followers proceeded from "the law of the excluded middle". That is, any statement is either true or false (the meadow is either *green*, or *not green*). In this sense, we wouldn't be mistaken, if we call the scheme **"either ... or..."** as the main construction material of Aristotle's logic. The meadow Aristotle saw, most probably might be not completely green in another season of the year. It might be little green, less green, green-like (how powerless the language is). Even in the late autumn we can see the last and only piece of green grass in the meadow. Thus, we witness how our potentials are induced by the scheme **"either ... or..."**.

If stated in the simplest way, the fuzzy logic proposed by L. Zadeh is a transition from the scheme **"either ... or..."** to the scheme **"neither ... nor..."**. Thus, it is proved that it is owing to the scheme **"neither ... nor..."** that the diversity in the objective reality finds its description through natural language more adequately. That is, the scheme **"neither ... nor..."** is more relevant to "the rhythm" of our thinking and provides the human being with a wider choice. Not only two poles, but also many (an infinite number of) variants between the two poles take part in the selection process along with these two poles. We can also call it "the world of quantums". An ancient man who made use of this mode of thinking strives with more enthusiasm to describe the diversity in life. He/she accelerated as much as possible (and sometimes artificially) the development of language at the expense of the energy of his/her internal psychoneurological potential. Yet, we know that it is simply impossible to carry it out adequately. Language will never describe the objective world as it is in reality. It will have no power to do it. Language will refuse from it "by pretending to be sick". The human being will never overcome his/her language in this connection.

On the other hand, the objective world itself would not show to the human being all the possible shades (quantums) for adequate self-understanding. This is far beyond the human potentials. Even Nietzsche's extra-ordinary could not have achieved it.

However, despite all these restrictions, human effort to get up from the Procrustean bed which has satistfied him/her until a certain moment in his/her mind, his/her initiative to see new, more than two shades and contents are noteworthy by themselves which should be accepted as a manifestation of his/her determinal development. This human being has already developed from crawling to standing up. New horizons are opened before the eyes of this human being. He can look much further and most important is that he can see that far. At this primitive point, language is capable of meeting its primitive challenges. Rapidly appear new words. So far pretending to be sick does not occur to language. This will happen a bit further when development will take place in-depth, not lengthwise. As the words acquire polysemi due to the lack descriptive potential, a path will be opened for metaphoric thinking, linear-informative mode of thinking and utterance will start giving way slowly to the mode of literary thinking and utterance. The language of invariants will turn into the language of variants. Certainly, "the arrival" of the invariants had occurred earlier than "the arrival" of variants. The poles forming the bipolar logical chain (the invariant model) were to sever from each other by means of an explosion so that they could shine together with all the shades (variants) to the quantum in their wombs.

A bit earlier we implied that Aristotle's formal logic does not allow the ancient man to choose more than two and thus, prevents him/her from seeing a complete picture of the diversity in objective world. The human being is laid onto a bipolar Procrustean bed built with the scheme **"either … or…"** in this way in the deepest layer of his/her mind. In this scenario, there is no talking of potential (yet, relative) diversity in language. Since, according to this formal logic, there are two colours: *white* and *black*.

The mindware sets this binarity in the poles *true* and *false*. Mind is obliged to get used to this binary contrast. We could say that such kind of practice has cost the mankind a lot. It is this bipolar ideological basis that stands in the soul of many injustices that have reigned all through history, all kinds of absolutes, the cases of allowing others to have what you did not consider it fit for yourself, to be more precise, the wars, avoiding alternatives. And, in its turn, all this preconditions the failure of democracy to come by its natural pace. While in one case democracy manifests itself, in another it fails to, and so on and so forth. The resistance to arrive in a systematic way leads to obtain democracy through a magnifying glass element by element from one or another source. We happen to look for it in a corner of some law or in some ruler's action or in poets' imagination. Or like in this book, we are searching for democracy in the language of an epos ("The Book of Dede Korkut") which has rooted prior to the ancient, mythological history, and this epos provides us lavishly with rich elements of the transition from absolute mindware and lifestyle to a democratic mode.

And finally, as we mentioned earlier, in 1965 the motto "irreconcilability with colorlessness" seemed to have guided the opponents of Aristotle's logic how to get out of the mystery. It provided the power of getting up from the Procrustean bed. The theoreticians of fuzzy logic started to notice the infinite color shades between the opposite poles the way the white and black TVs switched to the color regime. And we can state that Zadeh's works [4], which have started their circulation on this topic since that date, made a revolutionary coup.

And so, His Excellency Theory of Fuzzy Logic starts to penetrate into every sphere of our life following this revolutionary coup. The American-Azerbaijani scientist Lotfi Zadeh becomes the author of this breakthrough. The role of this outstanding scientist in the discovery and formation of fuzzy logic as a scientific domain is inimitable.

Here it is appropriate to remember the well-known physicist Albert Einstein's idea. In his book "Geometry and Experience" he writes, "So far as the laws of mathematics refer to reality, they are not certain, and as so far as they are certain, they do not refer to reality" [6]. According to the American scientist Bart Kosko, the mathematical world does not correspond with the real world described by it. These two worlds differ from each other. One of them is artificial, and the other is the real world [7]. From this point of view, we can state that the main force of fuzzy logic is its being more adequate.

Contrary to the scientific view established for thousands of years, L. Zadeh first put forward his theory of fuzzy logic as a hypothesis. Further, he proved that it is impossible to learn the complex processes taking place in nature and society, especially different cataclisms in the framework of the classic mathematics based on Aristotle's binary logic.

Developing his ideas which were original, but not accepted by "the Orthodoxes" at the beginning, L. Zadeh showed that the score of the truth of this or that statement can be defined not by two logical levels—1 (true) and 0 (false), but through the sets of infinite degrees (through quantums) between these levels. Thus, getting expanded owing to fuzzy logic and fuzzy set theory (L. Zadeh's theory), the mathematics based on binary—dual logic and set theory acquires a new content. This content manifests itself in the structure of the epos under our studies as well. We say "in the structure", since it exists not only in the language layer, but also in the content layer.

It is curious that these infinite scores degrees between "true" and "false" in the epos "The Book of Dede Korkut" manifest themselves also at the level of appraisal of the characters (at the level of content). The Epos has characters who enjoy two contents. One of them—some of them are "robotlike" characters pertinent to the ancient epos creativity, and it should be noted that they are in majority. This kind of characters are those who were formed by the principle "either… or…". They are either sympathetic or anti-sympathetic. For instance, Basat and Goggle-eye. It is possible to include one of them to a camp with a positive mark and another to a camp with a negative mark. In this sense, Salur Kazan, Kan Turali, Burla Khatun, Uruz, Begil, Yigenek, Boghach can be appraised as positive characters. They bear a psychological burden. They are unambiguious and mono-functional.

The other camp is settled by Goggle-eye, Yaltajuk, son of Yalanji, Aruz Koja with a Horse-Mouth, the princes of the infidels. They have one characterization—to create a negative content. They are also unambiguious and mono-functional.

It is specific that the characters in both the camps are united in terms of being unambiguious and mono-functional. Their psychological burdens are simple, their characterization does not recognize several semantic burdens. They are either good or bad. They are either true or false. The characters from both the camps can be fitted to one pole from this point of view according to Aristotle's logic. These characters are like robots. Basat is a monotonous character, he lacks psychological complexity. The same is true regarding Boghach, Kan Turali.

On the other hand, among the characters of the Text we encounter such ones who act in the bi-functional content. It should also be noted that this multifunctionalism in the Epos is just being created. For instance, Beyrek is such kind of character. One can also ascribe Goggle-eye to some degree to this pole. It is possible to see this quality in Salur Kazan.

These characters can be branded as new type of characters for an ancient epos. They bear not only a single psychological burden, but several. Various psychological loadings manifest themselves in them [1]. They are at the same time both good and bad. They are both loyal and cunning. They seem as if being formed to escape becoming a robot. Beyrek both maintains his loyalty to Salur Kazan, even meets his own death for that, also breaks his promise given to the daughter of an infidel, does not return to marry her. It is clearcut with Goggle-eye, however, he can also come to desire to commit suicide having been fed up with the violences he has committed. He experiences a genuine psychological dualization. Salur Kazan is the guard of the Oguz, but also some of his inadequate actions lead to many confrontations. For instance, the real reason for Outer Oghuz' hostility to the Inner Oghuz was Kazan's negligence towards the Outer Oghuz princes and their noninvolvement by him while he had his house ransacked. And it was Kazan who caused Begil's annoyance and deserting Oghuz. In the presence of the Oghuz princes Kazan appraised Begil's feat shooting an arrow to stick the deer's ear to its foot as the feat of the latter's horse.

In this way, new types of characters with psychological complexities appear in the text of the Epos. These characters begin to differ from the robot-characters for their nature. They give a more adequate reaction to the variety in the ontological world and try to represent it more adequately in their own examples. Psychological burdenings are various psychological shades acquiring their place between "true" and "false". We think that having these shades implies acting in the regime of fuzziness.

Through time the theory of fuzzy sets was practically implanted to all the sciences, including medicine, social sciences, sociology, conflictology, etc. [8].

The introduction of the fuzzy parameter of infinity in practical scientific studies enabled the consideration of infinity adequate to the processes taking place in nature, society, and in micro and macro worlds. Science approached the ontological level one more step. It is owing to this that the theory of fuzzy logic and fuzzy sets not only have found their dialectical approval, but also caused the formation of a new scientific direction, as well as in a broader sense, the formation and understanding of a new philosophical world outlook which is the next stage of cognitive theory.

Initially, fuzzy logic was mainly applied to technology and industry. Especially, in Japan, China, Korea, the United Kingdom and other countries, this logic was first steered towards industry. Serious financial profits were gained with its help. The first application was the fuzzy controller created under the scientific supervision of the authors' close friend and colleague, world-renowned British scientist E. Mamdani and is harnessed in the cement device today as well [9].

Further, fuzzy logic was applied to the fuzzy management of the trains running without drivers. And beginning from this, fuzzy logic found its wide application to the preparation of computer chips weighing all in all 5 g. If we state that the trade circulation of just this 5-g fuzzy chip on the world market is annually over 20 billion dollars, we would have indicated though partially the economic potential of fuzzy logic. The system with such a crystal can manage any fast process, even for example, the process of targeting of a zenith missile. One needs quite a little time to teach this chip some fuzzy functional skills like "fast-slow", "far-close" and to record the findings on the fuzzy regulatory crystal [10].

However, it should be pointed out that while creating this theory, L. Zadeh's main goal was to apply it mainly to social and humanitarian spheres, i.e. to the spheres that are mathematically unstructured rather than to technology and engineering. That was his dream to solve humanitarian problems surrounding human beings. He repeatedly talked about this dream of his to one of the co-authors of this book, Rafik Aliev whom he considered his friend and his follower in science.

We think that this book, i.e. the application of fuzzy logic to such an epos like "The Book of Dede Korkut", which puts forward the people's ancient history, their mindware will be a tiny contribution to the realization of this dream of the outstanding scientist L. Zadeh. Such a start can be a stimulus to discover the examples of fuzziness existing overtly or covertly also in other literary-cultural texts all through our history.

The discovery of many new horizons by this scientific direction and the chance of its first application to ancient literary texts also suggest us that its scientific potential is unlimited. It should especially be noted that in this line the epos "The Book of Dede Korkut" is our first literary text to be reviewed in the context of fuzzy logic. Its possibility is an indicator that the principles of fuzzy logic hold an important place in the mental and language corpus of ancient Oghuz stories. The possibility of viewing our literary texts in the light of fuzzy logic already suggests at least the antiquity of this text (its affiliation to mythological history). At the same time, this approach allows us to have a real idea how our language acts (or will act) in a given situation in that mythological history. It enables us to understand more clearly the interrelation between language and consciousness.

* * *

Right from the start, we should point out that fuzzy logic, first of all, is the logic of doubts, hesitations, and maybes. It is the logic of searches. It is the logic of efforts and endeavors. It is the logic of exploration and discovery of new horizons. It is the logic of a race towards quantums. This logic appears not unreasonably. Moreover, it appears with its hidden covert ideology which is lurking in its depth and maintaining the choice for all these "maybes" within itself (as if Athena Pallada had been born coming from the head of her father Zeus in her full clothing as a warrior) and thus, terminates the rule of binary confrontations, the absolute of invariants.

There are four reliable facets in this conception: (1) the set-theoretical facet; (2) the logical facet; (3) the epistemic facet; (4) the relational facet [11].

These facets shape fuzzy mindware proceeding from the system of various relations. Let us have a general look at these facets separately.

3.2 Brief Information on Facets of Fuzzy Logic in "Dede-Korkut"

The facet, which is called "The logical facet", is a generalized form of multifield logic. It spans like a bridge between "true" and "false", "the earth" and "the sky". This bridge has both the ends (poles), as well as an infinite and unbreakable path between both these ends. It should be reminded that contrary to this, in Aristotle's classical (formal) logic only the opposing ends (poles) were obvious. There was only the scheme **"either... or"** and the two poles originating from it!

The law of *"the excluded third"*, which serves as the main basis in classical logic, does not work in this facet of fuzzy logic, there is no space for it here. This law which belonged to Aristotle suggested that there can be no mid case between two opposing poles due to the excluded third. All the shades, variants, nuances were refused by Aristotle. To be more precise, let us formulate it this way: the imperative mindware did not allow this genius person to see or imagine them.

However, time proved that Aristotle's formal logic stating that there is only a gap between the scheme **"either… or…"** is not always true. Actually, the excluded third can also be excluded. That was what L. Zadeh proved. This is how one can call this case. The exception of the excluded third!

The presence of a third, a fourth (…) between the poles turned into a basis for this new theory of L. Zadeh. We clearly see its proof in our ancient epos "The Book of Dede Korkut". And throughout our study we'll call this case *"the exception of the excluded third"*, as we have mentioned it a bit earlier.

In ancient texts we encounter this case in the discourse of the sentences within the text, between the lines and in the hidden layer of the text. "The exception of the excluded third" throws us from the flow of a big and thick river making its way in the frame "**either … or…**" in our consciousness into an infinite sea which is being shaped through the scheme "**neither … nor …**". We see the blossoming, sprouting, birth and increase of invariants. Variants are being born. We at least imagine the depth of these variants, since the human being is not destined to see it [12].

We encounter the case of the exception of the excluded third in the epos "The Book of Dede Korkut" at the most distinct points. Here the characters set forth in search of other nuances existing between the two opposing poles, consequently, they actually manage at least to dip their feet into those waters on the shores of the ocean of infinity in the world. In our epos there is a third, forth, even it is not deprived of a fifth.

The main aim of the facet called "fuzzy sets" of fuzzy logic is that everything, every phenomenon in the world except the God has a degree. If we review a fuzzy set, then we accept that all the elements in the universe have a relation, degree of affiliation to this set.

The fuzzy sets are a solid means encompassing the restrictions imposed on linguistic information in the analysis of literary texts. Let us make such a note. When some incomplete information is provided in association with the action or phenomenon described in the text, this decreases the trust in it. For instance, the expression "certainly/evidently" is semantically stronger than the expression "probably". To make "probably" stronger than "evidently", one needs to add the word "most" to the beginning of "probably" and introduce the expression "most probably". One should take into account also the determiners like *"little"*, *"much"*, *"more"* in this connection. Their attitude to the text is defined by the force of their internal semantics. Also, the role of a context should not be forgotten either. The context polishes the mission of those determines a bit further.

It should be pointed out that this facet enables the assessment of uncertainties in the text. A bit earlier we mentioned that when some incomplete information is provided on any action or phenomenon in the text, this decreases the trust in the text. Thus, during the processing of linguistic information, we should pay a special attention to the factor of modality indicating how the action or phenomenon takes place or how it will be. The means serving modality express the degree of the receiver's trust by the sender and we can say that the expressions of modality also carry out the mission of softening the absolute "truth" of the said. And this vividly testifies that the existential essence is "at work". It can be accepted as a hint to the transition from "absolutism" to "democracy". We are going to speak about the softening (or sharpening) means of the text a bit further.

The epistemological facet of fuzzy logic is associated with the description of the knowledge introduced in the texts, the semantics of natural language, the analysis of the information in texts. In this aspect natural language is accepted as the description system of perceptive information, i.e. the information that cannot be measured directly. This facet is based on the theory of possibility rather than the theory of probability. Epistemological approach implies that the main goal is to be able to pull out of the hidden layer of the text and assess the information concealed by the author who is reluctant to reveal it (or is unable to say or reveal). In the epos "The Book of Dede Korkut" the points that the author is reluctant to announce openly or tries to conceal are not few. By defining them, we are obliged to descend to the depths, the hidden layers of the text. For instance, a hidden point associated with Beyrek in the epos can be revealed only by descending to the hidden layer of the epos, and the information in this hidden layer helps us to understand some imperceivable moments in the overt layer. It should also be pointed out that the author's intention to conceal completely should be distinguished from his effort to conceal partially. It seems to us that not always can we discover the point which the author intends to conceal. If the author does not provide us with a clue, there is no descending to the hidden layer of what he/she has created. There is such a clue in the epistemological point associated with Beyrek. The text presents it to us covertly.

In the description of Beyrek's life in captivity we encounter a curious moment. In the text it says [13],

> Məgər kafir bəginin bir bikr (bakirə) qızı vardı. Hər gün Beyrəyi görməyə gəlürdi. Ol gün yenə görmege geldi. Baxdı, gördi Beyrek səxt olmuş. Qız aydır: Neçin səxtsən, xanım yigit?! Gəldigimcə səni şən görərdim. Gülərdin, oynardın. Şimdi noldun?! dedi. [Now, the infidel lord had a maiden daughter, who loved Beyrek and came every day to visit him. That day again she came. Seeing Beyrek looking downcast, she said, 'Why are you downcast, my kingly warrior? Whenever I have come, I have seen you cheerful, smiling and dancing. What has happened now?!]

There is a reason for Beyrek's despair. He has received the news that his betrothed, the Lady Chichek, is going to marry another man. The news was delivered to him by his father's tradesmen. And the text gives us such a clue associated with quite another unknown aspect of the nature of Beyrek who was "cheerful", "talkative" while being in captivity. For a long time, the researchers have neglected this clue, they didn't "notice" it and didn't make use of it. Whereas, if we had not pulled down the curtain off the very important points in the overt layer of the epos, we couldn't have been able to discover the systematic nature of the chapters of the epos and observe the interrelation of the corresponding points in different chapters. This Beyrek wouldn't have helped us to know the other Beyrek. Also, if we hadn't used the clues, new types of characters (with psychological burdens) formed in the formation womb of the epos would have been left out of the focus of attention [14].

Actually, epistemological approach in a broad sense rises above the entire metaphoric space. Any metaphor means another way (a bit differently, quiet differently, almost 100% differently, etc.) of expressing by language a deeper and objective truth in the objective reality. Another way of expression cannot be complete or whole, it definitely gives way to distortions. And it was not in vain that our great poet Fuzuli hinted at it. Let us remember his line of great philosophical depth anew: "Don't get deceived as the poet's word is certainly a lie". It is a lie, that is, a distortion, isn't 100% reflection of what is in the objective reality. Since, our language is always lagging behind our consciousness.

Anything epistemological has a fragile status, which doesn't mean that it remains only secret and can be completely revealed only through perfect manipulations. As an object, it doesn't exist ready-made, can exist only as a conceptual side of the object (as an image in the brain).

To see Beyrek completely and as a whole, it is not enough for us just to learn through narration from the outer layer of the Epos his being faithful to Kazan as the favorite of the prince of princes. One needs to build the unity of this overt side of Beyrek's essence with his cheerful, sociable covert life in captivity (the life hidden for the reader). The complete and whole Beyrek is born out of the unity of these two poles. The Beyrek in captivity is also incomplete (fuzzy) by himself. So is the Beyrek who has been demolished owing to his faithfulness to Kazan. One should merge these "two" Beyreks in the mind's eye.

Thus, as is obvious, fuzzy epistemology helps to define the fuzziness in the text, draw away the fog of complexity and fill in the gap.

The fourth facet of fuzzy logic is the facet of fuzzy relations. This facet incorporates fuzzy relations, more precisely, the fuzzy interdependence between cause and effect. Fuzzy phrases (variables) and the relations based on the scheme *"If..., then..."* built on their grounds comprise the basics of this facet. We can point out the sentences tightly constructed by this pattern of thinking more often in the Introductory part of the Epos.

In the text it says,

"Əzəldən yazılmasa, qul başına qəza Gəlməz;

[If it is not written from all eternity, no disaster befalls any mortal's head;]

Ulaşıban sular daşsa, dəniz tolmaz;

[Though the rivers rage and overflow, the sea is not filled;]

Qara eşşək başına üyən ursan, qatır olmaz;

[Though you throw a bridle over the black ass's head he does not become a mule;]

Hecesindən düz oxunsa, yasin görklü...

[The Sura Ya Sin, when recited correctly, syllable by syllable, is beautiful…].

Yapa-yapa qarlar yağsa yaza qalmaz;

[Though the snow falls in huge flakes it does not last till summer;]

Qaravaşa don geyirsən, qadın olmaz;

[Though you dress a captive girl in a robe, she does not become a lady.]"

These real examples can easily be considered as the lexical-morphological derivatives of the scheme *"If..., then..."*.

The facet of fuzzy relations implies some close inter-relations of some elements of the sets reviewed in different situations.

Fuzzy relations are a very important notion in fuzzy theory, and they can be used in many spheres. As we have mentioned, this connection is realized in the scheme *"if..., then (at that time)..."*.

To describe fuzzy relations, we'll review a simple example we meet in our everyday-life by using, first of all, discrete fuzzy sets.

If **X** is the colour of the fruit and **Z** being its degree of turning ripe, then the relations between them can be described as follows,

Suppose, colour is given as the **X** = {green, yellow, red} of the fuzzy set, the degree of turning ripe **Z** = {the fruit is not ripe yet, is partly ripe, is completely ripe} is given for another fuzzy set **Z**.

In that case the relations between the two fuzzy sets can be described through the scheme "if…, then (at that time)…" as follows:

1. If the colour of the fruit is green, then the fruit has not ripened (it is unripe);
2. If the fruit is of yellow colour, the fruit is partly ripe;
3. If the colour of the fruit is red, the fruit is completely ripe.

In the chapter "Emran the Son of Begil" the text of our epos displays interesting points associated with this facet in the examples of the replacements of Begil and Emran in the enemy's eyes. This will be discussed in one of the forthcoming chapters.

* * *

We should underline again that the main advantage of fuzzy logic over Aristotle's formal logic is that it does not exclude a third, a fourth… between two opposing poles. It does not restrict human thinking. Thus, it turns out that our ancient ancestor's mindware was built on a democratic approach. Since, our ancestor is not confined to categorical (absolute) approach, he/she wants to see and describe the world in its diversity. He/she dives into hesitations; doubts do not leave him/her alone. Being armed with the very factor of doubt and its main component, the unit *"maybe"*, he/she even can see the transition stage between the *dead* and *alive*. By inserting a human existence into the scheme *"either ... or... "*, Aristotle considers the job done. Proceeding from his own experience, this great philosopher of Antiquity thinks, *"The man is either dead or alive. A third case is an exception."* This example has not left the textbooks of logic as an example of classic formal logic so far. However, when we review "The Book of Dede Korkut" through its covert layer, we can see that the mindware we encounter in our epos can provide new potentials to contradict it. Let us focus on a point. In the text it says,

> «Ol zamanlar Oğuz yigitləri yatanda yeddi gün, yeddi gecə yatardılar. Adına küçücük ölüm derlərdi.» [((In those days)) It seems… that the Oghuz nobles used to sleep for seven days on end. That is why they used to call it 'the little death'."].

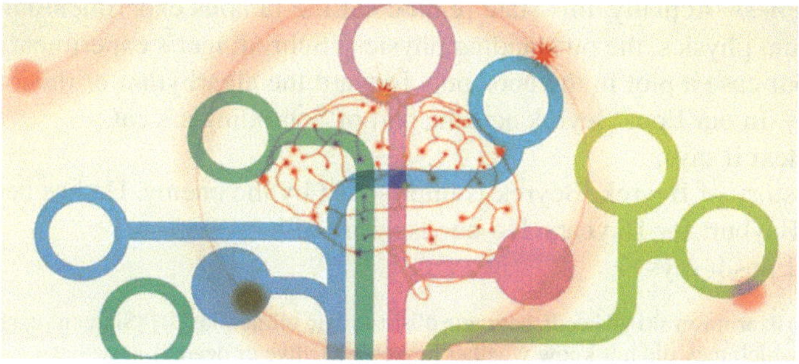

"The little death" is actually a state of a human being *neither dead, nor alive.* One can see in the epos that the characters sometimes find themselves in such a state. Kan Turali and Salur Kazan can be cited as examples. It becomes impossible "to wake them up" from the little death.

On his way back to the land of the Oghuz, with the yellow-dressed Seljan Khatun whose love he had earned having gone through different trials and thus whose father the bey of Trabzon gave his consent for their marriage, Kan Turali halted to have some rest. Seljan Khatun's father has already regretted and sent a punitive group after the two lovers to return his daughter. The troops of the infidel are already very close to them, whereas Kan Turali has fallen asleep a bit earlier. Seljan Khatun cannot wake hm up from the sleep called "little death" come what will. And then she is obliged to fight by herself against the punitive troops of her father's.

Kan Turali's finding himself in such a state and Salur Kazan's "little death" in another story is a moment utterly unexplainable for ancient man. As for modern men, this case is already known to them. Modern medicine calls this state "a **coma**" (a state between being dead or alive) [1]. Thus, by giving a name to this state, getting alieneated from the formal logic scheme *"either ... or... "*, contrary to it, our ancient ancestor managed to insert "the very law excluding the third", the basic point of the formal logic, into the frame of exception.

A new mindware—the one with the scheme *"neither ... nor..."*, which is closer to life and human beings, and describes the diversity in the world more profoundly—rises versus the mindware with the scheme *"either ... or..."*. Based on our example, we can say that the state of a **coma** ranks as "the third" between Aristotle's poles "*dead*" and "*alive*", hence a logical chain: *"Neither dead, nor alive. (If we recall von Neiman's 'maybe') He may be in a coma then?"* Thus, the place of "maybe" is further consolidated in our mind through the confirmation of fuzzy logic. As we can see, the scheme *"neither ... nor..."* lays grounds for *"maybe"*, whereas, the scheme *"either ... or..."* does not allow it. If formulated a bit metaphorically, *"either ... or..."* is the first sprout of absoluteness, while the scheme *"neither ... nor..."* is that of democracy.

And we sometimes encounter the one between *"neither ... nor..."* (the third one) in a completely different introduction. This introduction rises above the lack of knowledge. The knowledge of 50 per cent "dead" and 50% "alive" is delivered to the reader vaguely. And we can insert *a third inbetween the poles of the scheme of this vagueness.* Actually, this state reminds of the famous experiment known from the quantum physics, the outstanding physicist Schrodinger's experiment with a cat [15]. In our case a plot in the epos puts forward the algarhythm of this experiment artistically. In our Epos Beyrek acts the part of Schrodinger's cat.

In the text it says,

In the story of Beyrek, Beyrek is imprisoned by the enemy. He has been kept in castle of Bayburt for 16 years. No-one is aware of it.

In the Epos it says,

– Bunun üzərinə on alti yil keçdi. begrəgin ölisin-dirisin bilmədilər."– ["Sixteen years passed, during which they did not know whether beyrek was alive or dead."]

So, Beyrek is 50% dead and 50% alive for the Oghuz people. So was the cat in Schrodinger's experiment. After the explosion within the box where the cat was, for the onlookers it was not known whether the cat would remain alive or dead.

Finally, "an explosion" takes place in the Epos. Beyrek, about whom there has been no news whether he is dead or alive, returns to the land of the Oghuz, everyone hears and becomes aware of his being alive. However, it would be possible to add many moments inbetween the poles of Beyrek's scheme "dead or alive" until the explosion in the Epos. People were standing by either "the dead" side or "the alive" side of this 50%. And the presentation of Beyrek's shirt stained in blood to Khan by Yalanchi's son as a proof of his death was possible owing to Yalanchi's belief in Beyrek's death. If he had not believed it, he would not have told this lie so easily.

Meanwhile, Beyrek's father Prince Bay Büre is determined in the "alive" pole of the 50%. His mistrust in the news about Beyrek's "death", his doubtful approach to the news manifests itself in his sending the tradesmen in search of Beyrek even for the last time. Beyrek's maintenance in prison hidden from the Oghuz people facilitates the generation of relations between the belief in Beyrek's being alive and the belief in his being dead. Schrodinger also had hidden his cat from outsiders like this.

Thus, Beyrek known "neither dead nor alive" gives way to doubts, hesitations, searches, the search for him actually manifests itself as an expression of fuzzy logic mind.

Doubt can, in general, be reviewed as the constant of fuzzy logic. It is worth while tracing fuzzy logic where there is a doubt. Therefore, in the first chapter of this study we are going to introduce "constant doubts" of fuzzy logic. In general, we are of the opinion that it is possible to speak about the hidden or open presence of the moment of "doubt" in the natural language material which manifests itself in any facet of fuzzy logic. Moreover, it is important.

* * *

As we have pointed out above, sentences formed proceeding from the principles of fuzzy logic and constructed with the schemes *"the exception of the excluded third"* and *"if… then…"* are not few in the epos "The Book of Dede Korkut". The presence of these sentences shows that the constructive potentials of the creators of the epos are sufficiently immense owing to fuzzy logic. The degree of penetration of fuzzy logic into the language tissue of the epos is great. Moreover, this sometimes devastates the sentence frames and manifests itself on a wider description prism.

On the narrative prism bigger than a sentence "the exception of the excluded third" attracts the attention in the story of Boghach Khan Son of Dirse Khan. Let us focus on such a moment which is the soul of the tissue of the epos. The scholars in the Turkic studies have repeatedly addressed this point but from a different angle. For instance, according to this point, we can state that in this story of the Epos ("Boghach Khan Son of Dirse Khan"), which we've just mentioned, the situation associated with the ancient ritual can be clearly and correctly interpreted in its entirety, owing to the application of the principles of the very fuzzy logic. This example seems as if it were a live battle between Aristotle's logic and Zadeh's logic confronting each other in our ancient ancestor's mind. Simply, it should be mentioned that this battle takes place in a rather hidden depth.

Although the scholars in the Turkic studies underline the scene where Dirse Khan, who had neither sons nor daughters, was humiliated at the prince of princes Bayindir Khan's feast (being settled in the black tent) as an ancient ritual, they remained ignorant to another point—Dirse Khan's sufferings and the efforts to find out who was behind all this. Having left Bayindir Khan's gathering, Dirse Khan seeks the reason or the guilty for his being without sons or daughters and comes to such a conclusion after depressing thoughts: «*Bu qara eyib bana ya bəndəndir, ya xatundandır*». ["*This mysterious affliction of mine is due either to me or to my wife.*"].

Here Dirse Khan acts as a typical proponent of formal logic (Aristotle's logic). According to Dirse Khan, the logical chain consists of two poles—the poles "*me and wife*". The scheme **"either … or…"** is apparent. Whereas, there is a gap between the two poles—the poles *"me"* and *"wife"*.

It is according to this formal logic that the poles *"me–wife"* are shaped in Dirse Khan's mind. There is no any other reason in Dirse Khan's clear (!) logic. However, in the vision of Dirse Khan's wife—there is some. In between these formal poles, it is Dirse Khan's wife who brings the diversity, the inspiration to be much closer to life. According to her, the reason for being childless could be thought to be between these two poles as well. Dirse Khan's wife touches upon the exceptions. After being blamed by Dirse Khan, she presents to him some variants as the reasons for being childless:

1. Have you slaughtered of horses the stallions, of camels the males, of sheep the rams? You haven't. Have you filled the hungry, when you saw him? You haven't.
2. Have you clothed the naked, when you saw him. You haven't.
3. Have you saved the debtor from his debt. You haven't.

The wife, who recommended Dirse Khan to do all this, considers the lack of those efforts to be the reason for their being childless. The reasons stated by the wife manifest themselves as exceptions between the two poles. So, though the two poles of formal logic were the basics for Dirse Khan's search for the guilty (*me and wife*), his wife's introduction of different, various reasons between the two poles reveals a wider choice in the foundations of fuzzy logic. Thus, there is a case of alienation from the colorlessness of *"either me or wife"* (white and black) in the text. Exceptions are revealed. These exceptions include not filling the hungry, not clothing the naked, not saving the debtor from his debt.

The case of introducing "a third" (…) to the logical range and thus the entrance to the space of fuzziness manifests itself also in another part of the chapter.

Being convinced by the traitors who bring false news, during the hunting Dirse Khan has wounded his own son Boghach who remains wounded in the site of hunting. His mother casts about for him and reaches the foot of Kazilik Mountain. In the text it says,

> Dirsə xanın xatunu… qatlanmadı, qırq incə qızı boyına aldı. Bədəvi ata binüb oğlancuğın istəyü getdi. Qışda, yazda qarı-buzı ərinməyən Qazılıq tağına gəldi, çıqdı. Alçaqdan yuca yerlərə çapub çıqdı. Barsa görsə bir dərənin içinə qarğa-quzğun enər-çıqar, qonar-qalqar. Bədəvi atın öncələdi, ol tərəfə yüridi. – [Dirs' Khan's wife … could not bear it; she called her forty slender maidens to her side, she mounted her Arab horse and went in quest of her dear son. She reached Kazilik Mountain, whose ice and snow do not melt in winter or summer, and she climbed it. Up she galloped, from the low-lands to the high places. She looked and saw that crows and ravens were alighting in a valley and rising, settling and flying up. She spurred her Arab horse and rode in that direction].

This way, the mother finally finds her son, though wounded.

According to the mother, it is either the lion of Kazilik Mountain or the tiger of Kazilik Mountain to blame for her son's getting wounded. In her vision there is a scheme *"either a lion or a tiger"*. There is actually one more reason, a more realistic reason and that reason is brought to the chain of logic as an exception by the wounded Boghach. He says to his mother,

> Arslanla qaplanına qarğamagil. Qazlıq dağının suçu yoxdur. Qarğarsan babama qarğa – bu suç, bu günah babamdandır.
>
> [Curse not its lions and tigers;
>
> Kazilik Mountain is guiltless
>
> If you must curse, curse my father.
>
> This crime, this sin is my father's]

Boghach brings an important exception to the formal (clear) logic chain which reflects reality. With the involvement of the third (with the guilty father), a chain of fuzzy logic is constructed and thus, we witness the formation of a broader choice in mind.

* * *

"The exception of the excluded third" between the poles in the logical chain is brightly displayed also in another story of the Epos, the chapter "Wild Dumrul". Feeling pity on Wild Dumrul who asks the God to forgive his misbehaviour, the God gives such an instruction to Azrael so that *"Wild Dumrul should find a soul in place of his own, and his own soul can go free."*

A formal scheme *"either father–or mother"* is immediately constructed in the mind of Wild Dumrul who is encountered with such a challenge (to find a soul instead of his own). According to Wild Dumrul, either one of these people belonging to two poles must definitely sacrifice their soul instead of him. It is a strange logic, isn't it?!

An exception of the exception incompatible with formal logic fits itself between the poles *"either father–or mother"* which is a formal logical chain. Wild Dumrul's wife, a girl not of his tribe, is ready to sacrifice herself in place of Wild Dumrul. Thus, the scheme *"either.. or..."* makes way for the scheme *"neither... nor..."*. Neither father nor mother. Both of them avoid this grueling test. But who then? Wild Dumrul's wife, a girl not of his tribe, places herself between the poles by her own consent.

This point is an expression of another visual appearance of fuzzy logical thinking in the epos. Moreover, it is a bright expression. A little further there will be a more precise analysis about it.

* * *

A wider choice created by fuzzy logic in our ancestor's brain, first and foremost, suggests that the democratic status appearing with an extended choice is the basis for an initial attitude towards the surroundings and his/her own in our ancestor's thinking. Our ancestor wants to see the world and his/her close surroundings more colourful and with more numerous subtleties. He/she wants to make the right choice by increasing the very number of causes in the relationship **"cause and effect"**. To be more precise, this, actually, implies the winds of democracy blowing in our ancestor's mind. Even today the key point among the democratic principles is defined by a wider choice of the human being. In its language and expressions, the epos "The Book of Dede Korkut" has maintained up to present the incoincident, initial existence of these democratic embryos in our ancestor's mindware.

* * *

As a preliminary (!) outcome, we would like to point out this without waiting for the end of the study. While reading the book, the reader will encounter different manifestations of this preliminary outcome in its different places.

First and foremost, we should state that the text of the Epos shows vividly that a wider choice and the democratic approach which are the leading ideological principle of fuzzy logic, in other words, the spirit of tolerance has been pertinent to our nation since the ancient times, and this is an eternal value far from coincidence. Our ancestor learnt the democratic way of thinking by benefitting from the very principles of fuzzy logic. Their effectual use of these principles was possible owing to their inclination towards the width in their worldview, their interest in striving for the depths and subtleties of the problem. Consequently, various schemes were shaped in the brain, the nuances between the dual confrontations were revealed, the absolute of the invariants was ended, instead, the democracy of the variants commenced, the distinction between the situation existing in nature and the description of this state in mind and language was brought to a more (relatively) minimal state.

Certainly, we cannot say that one day this distinction will disappear completely. It is scary even to assume that. As such a state might be a sign of an absolute balance bringing the world to its end. Absolute balance means an end. Some distinction (even in the most minimal state) must manifest itself. At least, because the elimination of this distinction should always be a goal ahead. If treated from this prism, fuzzy logic is like a horizon, we would say. Owing to it, as humanity "discovers" the subtleties between the two poles one by one, the poles seem as if alienated a bit further from each other by themselves, however, this is a phantom, meanwhile quite new subtleties start twinkling in-between.

And we can build an urgent preliminary generaliziation promised by us a bit earlier as follows.

"The Book of Dede Korkut", the ancient epos of the Oghuz Turks, accomodates the principles of fuzzy logic in its linguistic space abundantly.

* * *

The mindware hidden in the depth of this ancient epos leads to the formation of the specific logical schemes pertinent to each of the four facets of fuzzy logic. And it turns out that our Epos has a sufficiently extensive language material on fuzzy logic, which itself implies that our ancient ancestor preferred a wider and deeper approach in his/her view of the world. This very approach enabling a wider choice becomes his preference. And a wider choice becomes possible only after the principles of fuzzy logic "start functioning".

The epos "The Book of Dede Korkut" embodies a democratic mindware which is manifested in the brain of our ancestor perhaps oblivious to it and can be considered the deep moral-ideological foundations for fuzzy logic. This mindware is not a concidence, it regulates our ancient ancestor's behavour pertinent to him/her since the most ancient times, his/her attitude to himself/herself and surroundings, and as a whole, vividly demonstrates the formation of our contemporary nation as a nation tolerant, loyal, multicultural, who cherish love in their soul, blood and mind towards everything surrounding him/her.

We are convinced that the methods and ways, principles suggested and substantiated in different chapters of the book will be useful in the appearance of new studies at least in the philological field.

The application of fuzzy logic to our monumental book "The Book of Dede Korkut" is providing this knowledge in initial approach.

We also find it appropriate to point out that our examples from "The Book of Dede Korkut" have been quoted from the source [13].

References

1. Abdulla, K.: The Semantic Twinkling in "The Book of Dede Korkut", p. 164. Mytholinguistics. Baku (2022)
2. Jaspers, C.: Meaning and meaning of history. In: C. Jasper Wrote About the Division of History into Two Periods—Prehistoric, and Historic Periods, p. 92 (1991)
3. Aliev, R.: Lotfi Zadeh is the genius of all the times. In: The Book "As Quiet and Strong as Water", p. 472. Baku, "Nurlar" (2019)
4. Zadeh, L.A.: Fuzzy sets. Inf. Control. **8**, 338–353 (1965)
5. "Plato's Middle Period Metaphysics and Epistemology" (2003). https://plato.stanford.edu/entries/plato-metaphysics/. Lukasiewicz, J.: Aristotle's Syllogistic. Clarendon Press, Oxford (1951)
6. Einstein, A.: Geometry and Experience. Prussian Academy of Sciences (1921)
7. Kosko, B.: Fuzzy thinking. The New Science of Fuzzy Logic, New York (1993)
8. Zadeh, L.A., Aliev, R.A.: Fuzzy Logic Theory and Applications. World Scientific, Part I and Part II (2019)
9. Mamdani, E.H., Assilian, S.: An experiment in linguistic synthesis with a fuzzy logic controller. Int. J. Man Mach. Stud. **7**, 1–13 (1975)
10. Shirinova, L.: Lotfi Zadeh—100, Baku, 608 p. (2022)
11. Zadeh, L.A.: Is there a need for fuzzy logic? Inf. Sci. **178**(13), 2751–2779 (2008)
12. Georg, C.: On a property of the concept of all real algebraic numbers. J. Renie Appl. Math. **77**, 258–262 (1874)
13. Zeynalovich, F., Alizadeh, S.: The Book of Dede Korkut. Baku (1988)
14. Abdulla, K.: From Myth to Writing, or Hidden Dede Korkut, p. 183. Baku (2009)
15. Schrodinger, E.: Die gegenwartige Situation in der Quantenmechanik. II. Naturwissenschaften **23**(49), 823–828 (1935)

Chapter 4
"The Construction" Material of Fuzziness in "The Book of Dede Korkut"

4.1 Hedges

The use of concrete language units, which are the manifestation of the principles of fuzzy logic, in the language of the epos "The Book of Dede Korkut" naturally induces us to analyse them separately. We are trying to descend to the depth of the Epos through these signs (we can as well call them the construction material of fuzziness). "The disclosure" of the secret layers of the epos in this way traces it back to the initial periods of the awakening of artistic thinking. Antiquity, the antiquity of the prehistoric mythological history in particular is animated before our eyes.

In the Epos some language units—separate words, reduplicative words, post-positions, numerals, indefinite pronouns, imitative words, adverbs, modal words—play as if a part of a construction material, as we have already stated, by developing the formal foundations for fuzziness. We'll call them hedges by one common name.

First of all, what is hedge and what do we imply by hedge?

We should point out that the theory of sets is based on the idea of linguistic variables [1]. A linguistic variable is a fuzzy variable. For instance, when stating, "Burla xatunun boyu hündürdür" ["Burla Khatun's height is tall"], this expression implies that the linguistic variable "height" is the linguistic value "tall". The range of possible values of the linguistic variable represent the universe of discourse of that variable. For instance, the universe of discourse of the linguistic variable "speed" can be between 50 and 220 km per hour and embraces such fuzzy sub-sets as very slow, average, fast, very fast. The linguistic variable preconditions the arrival of fuzzy set qualifiers—hedges following it. This way, through linguistic hedges the form of fuzzy sets changes.

As it is stated by L. Zadeh, linguistic expressions (if expressed mathematically, the variables) are such expressions that their values are formulated not by numbers, but with words and sentences. The representation of these values with hedges provides extensive potensials for research. For instance, a room temperature can be an example of linguistic variable. And its linguistic values can be expressed with the hedges

K. Abdulla and R. A. Aliev, *The Book of Dede Korkut and Fuzzy Logic*, Studies in Fuzziness and Soft Computing 438, https://doi.org/10.1007/978-3-032-05906-2_4

"cold", "hot", "high", "extremely much" (and an infinite number of others). The use of them adequately and appropriately to the point is associated with the context or the researcher's attitude.

The linguistic hedge is a means changing the meaning of a fuzzy set. For instance, let us answer the question, "How high can temperature be?" We can have answers to it like, very high, a bit high, less high, average high and so on. These are examples of various hedges which are the expression of the fuzzy set of temperature.

In real texts the following types of hedge are usually used.

- The hedges intensifying the fuzzy set (much, too much...),
- The hedges softening the fuzzy set (some, a kind of...),
- The hedges expressing suppositions (probably, it is not possible that...).
- The hedges expressing indefinite amounts (mostly, in rare cases...).

For instance, for the notion "Truth" we can use the following expressions (a complete false, plenty of false, a false, relatively true, true, very true, completely true). As we have mentioned above, this is associated with the use of the hedge to the point.

First of all, we should note that the hedges that appear on a language prism, in the text of "The Book of Dede Korkut" in particular, serve either to soften the degree of acuteness of the idea, content expressed or to sharpen the degree of its softness. This kind of "softenings" and "sharpenings" are realized both through real language elements and intonation. The words, expressions, reduplicative words, modal words... used in the Epos add the shade and spirit of modality to the context they belong to, thus serving to remove its acuteness or softness. In other words, we can say, owing to hedges, the text gets out of its frozen (still) state and movement appears on its surface.

It is more appropriate to consider the language units acting as hedges in the Epos separately. The main mission of hedges is to hide behind themselves the objective state, maybe or rather, to bring this state from "acuteness" to some "softness" regime (or vice versa). That is, the appearance of hedges on "the field" of the text is on the one hand, the result of the euphemistic process ongoing in the consciousness, on the other, this is indirectly associated with natural language potentials, is the use of its elements. In "The Book of Dede Korkut" we often encounter with hedges in the examples of various language elements. The euphemism in the consciousness is represented by various language elements (each time in a different degree). The outcome here is that hedges engulf the scene of objective reality in the veil of fuzziness.

Hedges seem to serve unveiling the cover. They provide the frozen state with motivation. So, they give us a chance to express more adequately the infinite shades of the real world which is the threshold of the world of fuzzy logic, we would say.

4.2 The Pronouns in the Function of a Hedge

If expressed in general, as it is clear from its name, the pronoun hints at another notion, reality lurking behind it. It is like a niqaab. The pronoun, as it is obvious from its name, is always replacing something. It has different types on language prism. Among these types the most interesting and important related to our topic is the personal pronoun "some" denoting indefiniteness. We think it appropriate to start a special talk on it.

The pronoun "**some**" (Let us not confuse it with the post-position "**some**"!) hides the value existing objectively in a text and concretely in the text of "Dede Korkut" and makes some impression about it. This way, the sign of fuzziness makes an impression about preciseness, we would say. So, the pronoun "some" makes the absence of clearness more prominent, as it is fulfilling the alienation mission from clearness.

It should be noted that as a hedge, more precisely, while fulfilling the function of a hedge, "some", in fact, clearly presents fuzzy mindware.

Let us focus on a point in the Epos. In the attire of a minstrel Beyrek comes to the wedding feast of his betrothed, the Lady Chichek. Beyrek himself doesn't want the princes to recognize him. Therefore, he has changed his clothes. He wants the Lady Chichek to recognize him first. Salur Kazan who likes this minstrel's shooting an arrow says to the princes around without knowing him,

> – Bəylər, bu günkü bəgligim bunın olsun. Qoyun nerəyə gedərsə getsün, neylərsə eyləsün.
>
> [Nobles! For today my realm is to be his. Let him go where he will, do what he will].

Beyrek does not miss such a chance and does what he will, adds to the wedding feast the elements of "a complete mischief". Finally, Salur Kazan gets a message, *"sultanım, dəli ozan, həp yeməgi dökdi, şimdi qızlar yanına varmaq istər." ["Lord, the crazy minstrel has upset all the food and now he wants to join the girls"].*

Salur Kazan allows "the crazy minstrel", i.e. Beyrek even to join the girls. As if Kazan guessed something intuitively.

In the text it says,

> Beyrək qalqdı, qızlar yanına vardı. Surnaçıları qovdu. **Kimini** dögdü, **kiminin** başını yardı. [Beyrek upped and went in the direction of the girls. He chased away the pipers. **Some** he beat, **some** he split the heads of].

Here it is the pronoun "**some**" or the hedge "some" that softens the state in the objective reality and therefore, is interesting to us. It is used here as a pronoun that denotes the indefinite number of people. This indefiniteness is indirectly the means of expressing fuzziness. That is, it is unknown how many of the pipers (which ones) were beaten and the heads of how many of them (which ones) were split of. And it is impossible to get information about it from the context either. Turning into a niqaab, again the hedge protects from us what is going on in the objective reality and hints at us so that we accept their absence (or presence). In this connection, let us focus on some other examples in the Epos:

– **Kimi** oğul deyü, **kimi** qardaş deyü ağladı; [**Some** mourned their son, **some** mourned their brother];

– **Kimi** aydır: keyik tozudur, **kimi** aydır: yağı tozudur; [**Some** said, 'It's the dust of a herd of deer'; **others** said, 'It is the dust of the enemy'].

– Oğlan qırx yigidin boyuna aldı, at dəpdi, cəng və savaş etdi, **kiminin** boynun urdu, **kimini** dutsaq elədi… (The boy took his forty brave men, challenged the enemy, and fought, **some** he beheaded, **some** he imprisoned"…).

– Qalın Oğuz bəglərin dəxi **kimini** zəbun edüb, **kimini** şəhid eylədi. (He left some of the entire Oghuz princes suffering and turned some to martyrs).

The pronoun **"some"** in these examples, being a hedge, set the linguistic elements around them into an environment of fuzziness. We do not know who is saying "son", how many they are, who is saying "brother", how many they are. Likewise, who mentions or mention "the dust of the deer", who is or are those mentioning the dust of the enemy?—It is not known. Who was beheaded and who was imprisoned?—It is not known either. The "author" of all these unknowns is the hedge ***"some"***. The special position of the hedge ***"some"*** as a pronoun, being a reliable, faithful soldier of fuzzy logic, is extremely typical for the text of "The Book of Dede Korkut".

One can also cite some other hedges with the function similar to that of the hedge ***"some"***. By spreading the fog of fuzziness, they enable the conditions bringing forth fuzzy mind. Let us focus on the following words cited from Father Korkut in the Introduction of the Epos.

Qorqud Ata ayıtdı: "Axır zamanda xanlıq gerü—Qayıya dəgə, kimsənə əllərindən almıya. Axır zaman olub qiyamət qopınca. *Bu dedigi Osman nəslidir. İştə sürülüb gide yürir… və dəxi **neçə buna bənzər** söz söylədi*". [*"Korkut Ata said, 'In time to come the sovereignty will again light on the Kayi and none shall take it from their hands until time stops and the resurrection dawns. This of which he spoke is the House of Osman and behold it continues yet.' And **many similar** things beside did he say"*].

"Many similar things ((words)) (=like this)" somehow expresses abstractly what Korkut said. *"Similar/of the like"* is not precise, it replaces something. One can only imagine what is behind it. And the ones that are hiding behind hint at the proceding content through the pronoun "this". Here is the shortcut to analysing the meaning of what is proceding and falling into the flow of meaning appropriate to them—"the word similar to this". Here the mode of thinking itself is fuzzy. Nothing concrete is said. This state is a secret in-depth way of avoiding *"white"* and *"black"* in the Epos. **"Of the like"** is not definite, since it is not "this in particular". Again we encounter a call for the independence of thinking. ***"Of the like"*** is a way of expression loyal to the main ideology of fuzzy logic—a wide choice. The presence of such an expression in "The Book of Dede Korkut" suggests the incampatibility of the mindware with the pattern of formal logic. Also, one should not forget that it circles the epistemological attitude as well, since it clearly hides all those not wished to be expressed within its pattern. The hedge "similar to this" plays a part of a real umbrella in this context. And we should not forget that hiding what has happened and describing them in another way is the basic mission of the epistemological facet, one of the facets of fuzzy logic.

In terms of its intellectual weight, the hedge "böylə" ("belə") [such] is the most relevant to the hedge "similar to this". On the language prism it is also a pronoun. The hedge "böylə" can accumulate as much informative load as possible in the text and thus, as if decreases the burden of the text, gives it a kind of "back-rub" (manipulation).

Dirse Khan's 40 unbrave warriors speak great evils in order to slander his son in his eye. They say,

> Sənin oğlun kür qopdı, ərcəl qopdı. Qırq yigidin boyuna aldı. Qalın Oğuzun üstinə yüriş etdi. Nə yerdə gözəl qopdısa, çəküb aldı. Ağ saqqallu qocanın ağzın sögdi. Ağ birçəklü qarının südin tutdı. [Your son... has turned out wicket and evel. He led out his forty young men and marched against the teeming Oghuz. Wherever a beautiful girl appeared, he dragged her off, re reviled white-beared elders, he tore the hair of white-haired women].

Having listed all these lies, the unbrave men continued as follows,

> Aqan turı sulardan xəbər keçə, arqurı yatan Ala tağdan təbər aşa. … Xanlar xanı Bayındıra xəbər vara. Dirsə xanın oğlı böylə bidət işləmiş deyələr. – [The news will cross the limpid flowing rivers, the tidings will climb the many-coloured mountain that lies askew, the news will reach the Great Khan Bayindir. 'Such and such monstrous evil has Dirse Khan's son wrought', they will say].

Here the hedge "böylə" draws the attention. "Böylə" actually puts all the lies previously told into its semantic "sack". It is a compressed, suggestive variant of the told lies—information. At the same time, it should also be stated that through this hedge the burden in the text gets decreased, it as if manipulated, as we mentioned above.

Another pronoun used in the function of a hedge in the Epos also draws the attention. This is the pronoun "all". It says in the text,

> Məgər qız meydanda bir köşk yapdırmışdı. Cəmi yanında olan qızlar al geymişlərdi. – [Now the girl had had a palace built on the square. All the maidens about her ore red];

The pronoun "all" is another ancient variant used in the language of the Epos. In modern language it is usually replaced by the pronoun "entire". Again this pronoun does not provide any definite information related to the number of the girls. Or let us consider such an example,

> Aslan hayqırdı. Meydanda nə qədər at varsa, qan qaşındı. – [The lion roared, and how much horse in the square pissed blood].

The hedge "how much" securely fulfills the function of a hedge and serves hiding the accurate information about the horses on the square. As we can see, the pronouns try to hide from us the objective reality as if pulling a veil over it. Yet, maybe a complete description of this objectivity is beyond the power of the language?!

Proceeding from the stated, it is possible to go further and reach an idea almost fantastic. Not only indefinite pronouns, even all the pronouns (their name reflects openly that they are replacing something), not only all the pronouns, even all the other words in the language are hiding from us the reality in the objective world like a niqaab. Let us recall Fizuli's immortal line, "Don't be deceived,as the poet's word is certainly a lie."

And while we are thinking like this, it is impossible not to see how the entire language confronts us like a huge, magnificent pronoun, like a hedge. So, one more function of the language is revealed. This is its function to hide the objective world from the speaker. We are confronted with such an outcome. The entire language corpus as a manifestation of fuzzy logic! It is worth thinking about this idea.

As for now, let us continue our talk about the other hedges in reality, in our case in the Epos.

Reduplicative Words as a Hedge

Diffuseness, uncertainty, the case of avoiding a clear-cut description and finally, fuzziness take place in the Epos through reduplicative words, which are the necessary component of the lexical structure. At this point reduplicative words act in the function of a hedge. The Epos presents abundantly the examples of reduplicative words acting as hedges. Let us consider this example used in many cases.

It says in the text,

> **Dönə-dönə** savaşayın, **dönə-dönə** vuruşayın. (Let us combat **again and again**, fight **over and over**).

As a sign of permanancy, reduplicative words (hedges) such as "over and over" are again hiding from us the clear side of the matter in the real world. It is as if thrown over all the battles like a veil. As a result, the visions are extremely diffuse. How many times did battles occur, haven't they differed from one another in terms of time, actors, space, to the ants and worms beneath that space? Certainly, taken into account the slightest differences, we are not going to meet a battle with a hundred percent identical moral and physical basis in any infinite span of time. The battles introduced to us in the infinity of "over and over" will always differ from one another and these infinite differences will become the moral basis of fuzziness.

The reduplicative word "over and over" is manifesting the alienation of the mindware from preciseness, in other words, from acuteness in its example. Or the use of concrete numbers in pairs again leads us to a space of fuzziness. While a separately-used numeral expresses preciseness and punctuality, the **reduplicative** variant of numerals, their usage next to each other lead to impreciseness and fuzziness. Let us consider such a moment of the story "Wild Dumrul".

While Wild Dumrul is following Azrael, who escapes him turning into a bird, he "*bir-iki göyərçin öldürdü, döndü, evinə gəli yürürkən...*". ["*returns home hunting one dove or two"].*

Let us focus on the hedge (reduplicative wor'd) "*one or two*". This is the formation of one fuzziness (*one or two*) by two separate distinctness (*one and two*) when they are together, next to each other. That is, the function of the word "*one or two*" is the formation of indefiniteness, fuzziness. Thus, as we notice, two distinctness next to one another generates one fuzziness.

If we consider other reduplicative words in the Epos, we can see that they are also bringing the air of fuzziness to the text. Let us focus on the examples,

– İçib-içib Ulaş oğlu Salur Qazanın alnına şərabın itisi çıqdı. [Salur Kazan son of Ulash drank deep, and the atrong wine rose up to his forehead].

– Qaza bənzər qızı-gəlini çığırışdılar... [his swan-like daughters and daughters-in-law screaming in chorus...]

– Ərənlər əvrəni Qaraçuq çoban sapanın ayasına taş qodı, atdı. Birin atanda ikisin-üçün yıqdı. İkisini atanda üçin-dördin yıqdı. [That dragon among warriors, Karajuk the shepherd, loaded his sling and shot. When he shot his first stone he toppled two or three of them]

– Dünya-aləm kafərin başına qaranqu oldı. [the world became dark about their heads].

– Gəli-gəli yurdının üzərinə gəldi. [He rode and rode until he reached home]

– Ölməgə-yitməgə getməmişdim... [I hadn't gone out to die or vanish for ever...]

– Evinə yazıdan-yabandan bir udlu qonaq gəlsə... [When a respected guest comes from the plain and the wilds...]

– Oğlan keyiki qovarkən babasının ögindən gəlib-gedərdi. [He was chasing and driving the stags and hamstringing them before his father].

– Quru-quru çaylara suçu saldım. (I brought water to the rivers that had run dry and dry).

– Tölə-tölə şahbaz atlarını bindilər. Qatar-qatar qızıl dəvələrini yetdilər. [They mounted his falcon-swift horses, stall after stall of them, they led behind them file upon file of his red camels].

– Ol qızın üç canavər qalınlığı-qaftanlığı vardı. (That lady had an attire three hurdles thick).

– Bin-bin ərdən yağı gördümsə, oyunum dedim. (Whenever I saw thousands and thousands enemies, I called it my game.).

One can state firmly that the reduplicative words spread to the text of the Epos like stars in these examples disseminate the mode of fuzziness to their surroundings. In each of them one can see the manifestation of the mind of indefiniteness and naturally fuzziness. In these examples there is continuance, infinity and also an attempt "to hide" the real scene. This function of hedges is fulfilled with great "valour" by reduplicative words used in the Epos.

4.3 Numerals in the Function of a Hedge

Sometimes to soften the thickness, acuteness in the text, to generate a peculiar euphemistic mood, numerals are also used. Let us consider the examples,

– Ağ alnında beş kəlmə dua qıldıq, qəbul olsın. [On your white forehead I invoke five words of blessing; may they be accepted].

– Otuz bin ər yağı gördümsə, ota saydım...(Whenever I saw thirty thousand enemies, it seemed to me like grass).

– Ala sayvan gög yüzinə aşanmışdı. Bin yerdə ipək xaliçəsi döçətmişdi. [his many-coloured tents reared up to the face of the sky. In a thousand places silken rugs were spread].

The numerals in these examples are formally indicated as real, but by essence they are indefinite. And like a niqaab, they are hiding in the background what is lasting longer and the number. In these examples, neither the performed prayer consists of five words, nor when speaking of thirty thousand troops, Salur Kazan has counted the enemy one by one, nor Bayindir Khan has silk carpets laid definitely in thousand places. What is hiding behind the real number—we won't know that. Since, we have already entered the zone of fuzziness (maybe a space which incorporated infinity). Thus, the Epos introduces the real elements of fuzziness and consequently, the logic of fuzziness also through simple numbers.

The hedges, which we have considered as "the construction" material of fuzzy logic, generate a softening (sharpening) mood in the text of the Epos. Thus, a protest is observed against the values of firmness, intolerance, steadiness, in other words, invariant absolutism. These waves of protest ongoing in mind are supported by the elements of fuzziness. The euphemistic nuances taking place in our ancestor's mind start developing through hedges which are "the construction" material of fuzziness and this, consequently, leads to the formation of democratic mindware and tolerance.

Reference

1. Aliev, R.A., Aliev, R.R.: Soft Computing and Its Applications. World Scientific (2001)

Chapter 5
The Text of "The Book of Dede Korkut" and the Constant "Doubt" of Fuzzy Logic

5.1 "Doubt" in "Dede Korkut" from Fuzzy Logic Point of View

First, one should say about the text of "Dede Korkut" that it is an anachronic text [1]. For its philological and humanitarian essence, this text cannot be ascribed to some certain (concrete) period as a literary output. This text includes "the tokens" of language and thinking of the prehistoric mythological and historical periods parts of which cross each other's borders. This text contains also the echo (a weak voice) of the period when a primitive man lived within Nature and considered himself/herself part of it. Likewise, it also contains a rich voice of the period (the period of "Culture") when the man of the period of Culture became distant from Nature and carried out almost the function of supervision.

It must be stated that the text of "The Book of Dede Korkut" records even the literary-artistic model of the transition moment from Nature to Culture. This fact itself is an obvious hint at what kind of antiquity the Epos originated from while being shaped. Such a suggestive model is "opened" by Basat's return to the forest (to dear Nature) after having been brought from the den of a lion in the forest to his father Aruz Goja's home (in other words, from the Nature, in whose bosom he had grown, to the Culture unknown to him yet). This return, actually, "the escape" implies that the transition to Culture has been achieved not so easily and that it has been painful and bitter. It is also curious that Dede Korkut himself takes part in this story as a character and plays a key role in severing Basat from Nature once and for all to enter Society [2, 3].

On the other hand, Dede Korkut records the events taking place closer to the period of the Prophet Muhammed (the 6th, 7th centuries) in the text, and Dede Korkut performs his function of an author in this nearest period as well.

Fuzzy logic provides us with the chance to both differentiate the echo as well as the rich voice of these two splendid periods and hear them.

K. Abdulla and R. A. Aliev, *The Book of Dede Korkut and Fuzzy Logic*, Studies in Fuzziness and Soft Computing 438, https://doi.org/10.1007/978-3-032-05906-2_5

Then a question arises. How can it be that as a personality, as an author, Dede Korkut is both within the events dating prior to those eras, he illuminates them and also presents those taking place in the period close to us? Is he a personality or should he be accepted as a principle?! The fact is that, if we approach the issue realistically, we then have to accept Dede Korkut as a principle, a mode of behavior, a controller of life and law, the way the culturological scholar Viko and philosopher Schelling stated about Homer. The Italian culturological scholar J. Viko and German philosopher Schelling accepted Homer as the name of the lifestyle and principle experienced in the Greek cities as a whole [4]. What is bad about accepting Dede Korkut actually not as a personality, but a principle?!

If viewed from the philological and humanitarian angle, "the logic" of fuzzy logic is born from the ideology of hesitations, doubts, assumptions, even somewhere the lack of trust. The great Azerbaijani poet Huseyn Javid wrote,

Şübhədir hər həqiqətin anası,
Şübhədir əhli-hikmətin atası.
Şübhə etməkdə həqlidir insan...
[*Doubt is the mother of every truth,*
Doubt is the father of the wise.
Right is a human being to doubt].

There it is: the soul of fuzzy logic. Actually, these lines can be considered the anthem of the ideology of fuzzy logic. During any simple or complex disclosure of fuzzy logic it is doubt that appears first. It stands at the outset of the journey of fuzzy logic. The component "doubt" is the constant indicator of the fuzzy logic compositions. It can be called just like this: the "Doubt" constant! It should be noted that in mathematics and physics the word "constant" is used instead of the word "daimi". For instance, we would cite the expression "Planck's constant" [5].

Only in case there is a "doubt constant", one can set forth to look for it in the labyrinths of fuzzy logic. The wise pointed out that doubt is not weakness in science and art; on the contrary, it is strength. Perfect scientific achievements are usually originated from doubtful hypotheses. According to Engels, doubt is the basics of scientific cognition [6]. And on the grounds of Kant's assertion, there are two things that evoke interest, doubt and respect in his wits: the Heavens shining like a star above him and the law of ethics within him [7].

While the main provision of Aristotle's logic (formal logic) is based on the ideology "either ... or", the doubt constant is oriented at the destruction of this falcrum.

The main function of "the doubt constant" is to guide the mindware from the scheme "either... or..." to the scheme "neither... nor..." The scheme "neither ... nor..." is an indicator of the transition of the mindware to the regime of democratization. Since, in this case we encounter the enlargement of the choice. Unlike the scheme "either ... or ...", the scheme "neither ... nor..." can definitely incorporate a third, a fourth... This is the expression of a more democratic mindware. Thus, while exploring the spirit and presence of fuzzy logic in the epos "The Book of Dede

Korkut" throughout this study, and finding the examples of it, we also happen to acknowledge the transition from authoritarian mindware to democratic mindware.

* * *

Fuzzy logic rejects everything considered absolute except the God. It says that there is no white colour in an absolute sense, there is no black colour in an absolute sense, there is no zero in an absolute sense, there is no one single in an absolute sense, there is no lie in an absolute sense and there is no truth in an absolute sense. If we transfer it to a social environment, hence is the root of unfairness and ungroundlessness of extremism, absolutism, totalitarianism. It is that time—when we do not have an absolute value around us—that doubts, hesitations start, the human brain dives into the search of it (of an absolute value) and during this search tries to see the variety, shades in the world.

The simplest definition characterizing fuzzy logic is stating that it is the logic of approximate reasoning. The distinctive features of the logic of approximate thinking are the following:

(1) the score of interpolary truth expressed by linguistic means of expression;
(2) the table of the score;
(3) the rules of approximate reasoning [8].

On conceptual bases such serious research naturally leads to revolutionary changes in science, technology and even fine arts. That is, fuzzy logic is, practically, "implanted" in all natural sciences, medicine, social sciences, sociology, conflictology. In practical scientific research the involvement of the fuzzy measure of uncertainty (we have reviewed it a bit earlier as "the doubt constant" in application to a literary text) enables the consideration of fuzziness which is adequate to the processes existing and taking place in nature, society, micro and macro world. Therefore, fuzzy logic results in the foundation of a new scientific direction which has found its dialectical proof, as well as the formation and understanding of a new philosophical worldview which is a subsequent step of the theory of cognition in a broader sense.

The charm of the theory of fuzzy logic consists of the abundance of mathematical solutions in the description and application process of the subject of cognition, as well as the presentation of different answer variants appropriate to concrete situations. The abundance of solutions in any description—research process implies automatically the existence of "the doubt component". This feature is one of the important qualities of post-neoclassic epistemology and the expression of non-unity necessary in the creation of theoretical systems. As a result of the day-by-day consolidation and strong effect (as a result of doubt) of the principle of non-unity, i.e. the idea of multiple solutions in scientific thought, such revolutionary qualitative changes take place in the doctrine of existence that stimulates the formation of new ontological systems [9].

Fuzzy logic is a fundamental discipline which has influenced not only scientific and philosophical view, but also the technical and technological development, social practice, the economic system of the world and will further do so. The role and merits of fuzzy logic are undeniable in the formation of modern information community, in the progress of information communication technology, in the development and management of intellectual systems, in the organization of modern education and development process on completely new bases.

* * *

"The doubt constant", which has a special function in the realization of the principles of fuzzy logic, has its own determiners (localizations). The most expressive among them is the component *"maybe"*. It is the determiner with the most precise border. It clearly and sharply reveals the idea, intention, that is, hesitation, disbelief. The components which manifest themselves as other determiners semantically close to it, for instance, *"it is probable that"*, *"fancy that"* and some others appear more *"softly"*. The determiner *"maybe"* is the most appropriate and the most "reliable" means in this semantic field.

The appraisal of uncertainties in the text is very important for the identification of the ideas in the substrata. That is, there may be some uncertainties that can influence the trustworthiness of the sense expressed in the text. Also, during the linguistic appraisal of the text, we should concentrate on the modality of the verb denoting how the action occurred or will occur. In general, the text provides some hidden information on the author's epistemic position based on his knowledge, observation or beliefs. In this case uncertainty is mainly expressed by using linguistic determiners.

All these determiners are associated with language facts, actually with "the doubt constant". It is certainly possible to encounter these expressions, words in either overt or covert layers of the text of "The Book of Dede Korkut". If in future studies there are efforts to find out these very determiners in the text of "The Book of Dede Korkut", we would be able to put the problem forward more profoundly.

It should also be noted that the search for the hedges (softening, sharpening words and expressions) in other literary texts is like an effort of seeing the particles of light of fuzzy logic in the dark. By seeing them we can first fix the fuzzy points in our literary text, and then, naturally, comment on them. We are convinced that such kind of studies will appear regarding other literary texts as well. For now we are hopeful that the presented book dealing with the study of the principles of fuzziness in "The Book of Dede Korkut" can be theoretical and methodological foundations for the mentioned studies in future.

* * *

5.2 "The Little Death" in "Dede-Korkut"

It is true that the text of "The Book of Dede Korkut" is a text which does not allow hesitations, doubts in its overt (open) layer as a whole. This is actually due to the general mood, ideology of the text. The clearly expressed presentation of the text has, as a whole, an intonation, spirit which does not allow to stumble, is away from poli-semanticism, and advances along the path of complete independence. However, this does not mean that there is absolutely no room for doubt in the text of "The Book of Dede Korkut". There is doubt, hesitation, but it is not overt, it is lurking in the depth. It is possible to bring it out to the surface by destroying some stereotypes. A doubt about Aruz Koja, a doubt about Goggle-eye, a doubt about Salur Kazan, a doubt about Karajuk the shepherd, a doubt about Yalanchi... Motives in the epos are like a pulled back bow not often revealing itself easily but kept in the hidden layer.

For example, some lexical components of the text lead us to the description of the perception of relations pertinent to the ancient period. By having a white tent pitched in one place, a red tent in another and a black tent on the third spot, Bayindir Khan adds a shade of red to the white (satisfactory) and black (unsatisfactory) variants. By this he displays the existence of a third (those who have neither sons nor daughters) between two poles (the pole of those who have no sons and the pole of those who have no daughters). This itself (distinguishing between female and male children of those who have children) implies the inspiration and attempt to see the world in a broader spectrum (in the diversity existing in it).

Let us address another point. In the depth of the expression *"This man is either alive or dead",* there is an implication of a doubt such as *"This man is neither dead nor alive, he may be in the middle of these two realities."* In the text of the Epos this supposition finds itself in the expression "little death". Wasn't this the name of the state of coma in that farthest period?

We have already pointed out that classical logic has the law of "the excluded third". This law says that if there are two contradicting ideas about one phenomenon at the same time, it is impossible that both are wrong. One of them is a lie, another one is true. Whereas, a third is excluded. Unlike this, fuzzy logic does not have the law of "the excluded third", here a third (fourth, fifth...) provision can appear (as a result of the doubt) between two contradicting ideas.

In the text of "The Book of Dede Korkut" this confrontation is enveloped in such a narrative adornment. There is a time span mentioned as "**ol zamanlar**" [in those days] in the text. This is a span of a more ancient period recalled with a nostalgic feeling. "In those days" everything was better, sons would not go against their fathers' words, girls would not appear in the presence of their betrothed and so on.

Such kind of "golden period" languishes in the mythological texts of many nations as the past that has slipped away and is mentioned with nostalgia. One can recall "the golden period" in Greek mythology. In the epos "The Book of Dede Korkut" such a revelation—the confrontation of the periods (*ol zamanlar—indiki zaman …*) [*in those days—the present time...*] acts as a fulcrum for fuzzy mindware to appear.

"In those days" it was not like this. If not like this, how was it then? The text expresses very greedily what it was like in those days. We almost have to imagine it.

"In those days..." A complete period of heroism! The nature, spirit, physical strength of giants would not match with those now (with the people in the epos). The confrontation between the moment in Nature and the moment in Culture! They are uncomparable!

In the text it says that "**in those days"** when the Oghuz nobles fell asleep, they used to sleep seven days on end, when it was impossible to wake the Oghuz nobles up by no means. Salur Kazan and Kan Turali had "experienced" such a state. This state should be called. Natural language has the power to do it. This state is called "the little death" in the text. There is neither real life, nor real death. The reality in the mid is the third, i.e. an exception.

Modern medicine also confirms the case of "the little death". The scientific term describing that state today is **"a coma".** Again fuzzy logic comes to the rescue and removes the uselessness of the law of "the excluded third". Neither dead, nor alive—maybe he is in coma?! Here doubt takes us to the threshold of the truth.

The name that appears as a result of fuzzy logic in the text of "The Book of Dede Korkut" *("the little death"—a coma)* proves that the name *"the little death"* was discovered owing to doubt, they refused to approach the man neither dead, nor alive with Aristotle's logic, there appeared a need to appraise this state by a new principle and consequently, such a name for "a coma" was discovered in the ancient Oghuz community and naturally in the Old Azerbaijani Turkic ("*the little death*"). First, the reality itself was observed (certainly, with the application of the element "doubt"!), then it was named. This name carried out the function of the third which is important for fuzzy logic.

A human being who lives in the period of Nature engraves the picture of a more colourful world with such a logic before our eyes. This is already the picture of the world whose author he/she is. He/she shows us a world with its own first colour and fragrance between "either... or...". These are the colours of the world that the human being saw (was able to see) while he/she was living in the period of nature. Unfortunately, the human being, who was nurtured in the bosom of Nature, has been losing (actually, forgetting) the function of seeing these colours, living with them and in them ever since he/she entered the period of Culture.

To apply the principles of fuzzy logic systematically to the text of "Dede Korkut" (actually, any literary text) is a very challenging issue. It is certainly possible that the random points that attract our attention may increase in number and sometimes will create a systematic picture. However, above all, it is necessary that the words and expressions within the capacity of our language, which are "asleep" or have escaped our attention, be trapped in "the net" of "the doubt constant" of fuzzy logic.

Harmonious philosophical ideas and shades find their manifestation through the formulas written with mathematical symbols on the one hand, and in poetic language on the other. It should be noted that the potentials of natural language are enormous. If natural language had had enough power when mathematics was created (let us recall "the disease of language"!), then physics, economics, social life, in general, the world we live in would not have been written with the present idealized formulas. The lack of power of natural language ("the disease of language") had led to the impossibility of naming the newly-created notions at the time when they were created. These names were created at the expense of the words already existing in the language which was powerless for the time being. It is here that one should seek for the genuine reason for the generation of metaphors. Also, "the disease of language" (its lack of power) made it lag behind the development of thinking. Or vice versa. The language's lagging behind thinking caused its disease. In such cases the language failed to turn into an adequate description mechanism of the diversity in the world or found it challenging to turn into one.

In this regard, there are great scientific perspectives for linguists. That is, through the natural language, which has restored its power and "recovered", we can describe almost any information. Through mathematical linguistics we can carry out automatic translation, modelization of creative processes, including setting in order the tales, stories, eposes, melodies and solve other issues. Whereas, we can conduct estimation, drawing logical conclusions in natural language. The key to the solution of the problems we have particularized is again fuzzy logic.

The human beings who communicate using natural language or any language have their own semantics, pragmatics to express their feelings, statements, behaviours. The human brain is a miraculous machine capable of turning the text /sound into natural language and understanding it. One should also take into account the presence of indefinite, fuzzy, unclear/ambigious and indescriptive features of natural language. Uncertainties can be solved through fuzzy logic. By understanding the presence of unclearness and uncertainty, fuzzy logic can provide the system processing the natural language with useful ideas. In the theory of fuzzy sets the appraisal process of the text in natural language (for instance, in the language of "The Book of Dede Korkut") covers three basic steps:

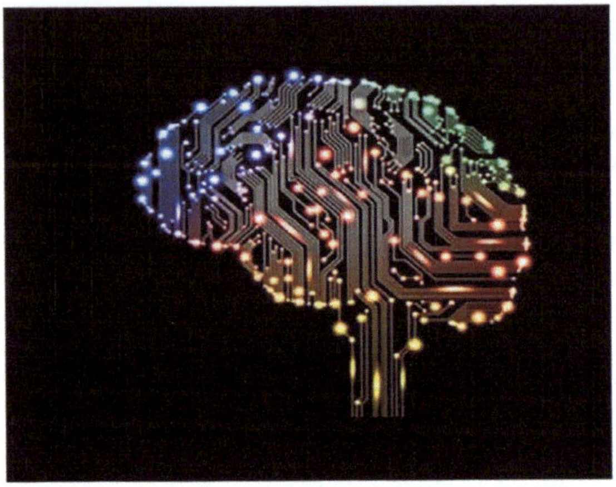

1. Dividing the text into categories to be studied (the division of the text into categories);
2. Prioritization or repeated points (the usage frequency);
3. Discovery of a special topic (of research).

Each of them makes their own "contribution" to the appraisal of the text, also realizes actually this appraisal.

* * *

It should also be noted that it is difficult not to agree with the view that contradicts and does not confirm the idea expressed by some scientists that "fuzzy logic will enter our lives yet from now on." That is, we would like to underline that the spirit of fuzzy logic had existed in any of our ancient literary texts, including, first of all, "The Book of Dede Korkut", in other words, in the "soul" of natural language since long. That is, we could as well say that even our ancestors were thinking in terms of fuzzy logic principles back in the prehistoric mythological period. The doubt about and hesitation towards his/her actions and statements manifest themselves in the lifestyle of our ancient ancestor like that of modern man. "The Book of Dede Korkut" is a literary-artistic space where these doubts and hesitations are most often encountered. The Epos is a favourable precondition for fuzzy thinking and hence for fuzzy logic.

We are of the opinion that fuzzy logic is rather the function of the period of "Nature". Either it has become fossilized, or been forgotten, and even if only forgotten, it has survived in our subconsciousness. Our current approach to it is our attempt to reinstate the ancient colourfulness. Sometimes it works, but sometimes it doesn't. When it doesn't, we'd rather say: it does not work for now.

References

1. Abdulla, K.: An Introduction to the Poetics of the Book of Dede Korkut, p. 17. Ötüken, İstanbul (2020)
2. Abdulla, K.: From Myth to Writing, or Hidden Dede Korkut, p. 183. Baku (2009)
3. Zeynalovich, F., Alizadeh, S.: The Book of Dede Korkut. Baku (1988)
4. Schelling, F.B.: Philosophy of mythology. In: Introduction to the Philosophy of Mythology, vol. 53, p. 21. St. Petersburg (2013).
5. "Planck's constant" haqqında. https://www.sciencedirect.com/topics/engineering/plancks-constant
6. Engels, F.: Quotes. https://kidadl.com/quotes/friedrich-engels-quotes
7. Kant, I.: Quotes About Ethics. https://www.azquotes.com/author/7722-Immanuel_Kant/tag/ethics
8. In our opinion, discourse is not a repetition of the textual elements this way or another. Discourse is a basis of additional knowledge the text brings (most often secretly) with itself
9. Aliev, R.A., Tarassov, V.B.: The man who changed the scientific world: to the centenary of the birth of Lotfi Zadeh. Adv. Intell. Syst. Comput. **1306**, 148–164 (2021)

Chapter 6
The Text of "The Book of Dede Korkut" and "The Search for the Third" Through Fuzzy Logic

6.1 Law of Excluded Middle in "Dede Korkut"

At times, when speaking of fuzzy logic, some people mean paradoxical words, the fine expression of contradictory ideas, the points incomprehensible through simple logic or obscure. This is completely wrong. Actually, fuzzy logic appears by noticing the manifestations of formal logic scattered over a wider "area", and its original, most primary function is to see the world not in two colors: white–black, but along with those between white and black. It is an attempt to trace the harmony of the process in mind, to view our language from this angle and remove the incompatibility in the middle (between our language and our mind). There is no doubt about the scale of the effort. The other colours potentially existing between the poles *"white"* and *"black"* (whitish, white-like, quite white..., blackish, dark black, super black, black-like...), shades (greyish), moments (bright, dark...) suggest that the world, relations are deep**er** and **more** colourful. Fuzzy logic is involved in discovery and presentation of those lurking behind the word "more". It carries out the mission of "whittling off" the incompatability between our mind and our language as far as possible.

Once this incompatability was described also by the great 16th-century Azerbaijani poet Fuzuli. His line presented in extremely deep literary generalization actually hints at the position of mind and language facing each other. The poet who says "Don't be deceived, the poet's word is certainly a lie" means that if compared with mind, it is a reality that unlike thinking, the word is a lie, i.e. not true, i.e. far from the truth, i.e. not corresponding our thinking. The attitude towards the loss when thinking was guided from the mouth cavity to this side (information channel) was also expressed much later following the great Azerbaijani poet Fuzuli, by the outstanding 19th-century Russian poet F. Tyutchev as follows, "...the idea uttered is also actually a lie..."

K. Abdulla and R. A. Aliev, *The Book of Dede Korkut and Fuzzy Logic*, Studies in Fuzziness and Soft Computing 438, https://doi.org/10.1007/978-3-032-05906-2_6

From the outset we'd better remind that in the introductory part of this book we have already started a general conversation about the existence of the elements of fuzzy logic in the text of "The Book of Dede Korkut". And here we would like to mention that in this and also forthcoming chapters we intend to embody a systematic view of the world of "Dede Korkut" in the light of the theory of fuzzy logic.

Presently, albeit briefly, let us have a glimpse of the fundamental structure of the theory of fuzzy logic. By doing so, we would have learnt also the consistency, effectiveness of the transition mechanizm from fuzzy theory to some certain literary-artistic text (in our case the text of "The Book of Dede Korkut").

In scientific literature, fuzzy logic is understood in two senses:

(1) in a narrow, purely logical sense;
(2) in a broad sense, in other words, in the sense of fuzzy sets and their clustered branches.

Logistical facet of fuzzy logic bridges the gap between "true" and "false", between "the earth" and "heaven". This bridge has ends (poles), as well as an infinite and continuous path between these ends [1].

In Aristotle's classical logic only the ends (poles) are obvious. In this logic of two ends the concept of "truth" of the binary logic begins "to work" and at this moment we encounter the manifestation of four main laws. One of them, **the law of identity**, is shaped as "Truth is truth" or "The text of "Dede Korkut" is the text of Dede Korkut". In linguistics it is also called "the sentences of identity".

The law of contradiction is shaped as "Truth is not untruth", or "Kazan Bey is not the one who is not Kazan Bey".

The law of excluded middle states that there is nothing between either the truth or not the truth, or either life or death. However, we already know that a third manifests itself between death and life in the text of the Epos. And this third is described in the text of "The Book of Dede Korkut" quite adequately.

It would be appropriate to remember here the expression "the little death" associated with Kazan Khan and Kan Turali, the reasons suggested by Dirse Khan's wife for lack barrenness, the brave action by Wild Dumrul's wife who was from an alien tribe and some other relevant moments.

And, finally, **the law of sufficient reason** is shaped by drawing from any text a general outcome based on its preliminary constituents. In other words, if we are armed with the law of drawing a rational conclusion, proceeding from the Introduction of the text of "The Book of Dede Korkut", we can grasp and appreciate the general spirit of the Epos.

It should be noted that these laws "work" only when the position is limited to the selection of one of the two possible points (poles), i.e. to Aristotle's approach. In this case we encounter ardour—the totalitarianism of natural mind: this is what it is—either this one or another one, there is nothing else! As we can see, the choice is minimal.

The mindware, which has gone through "Occam's razor", actually defines the boundaries of thinking. Any boundary is actually a Procrustean bed and implies the totalitarian essence of thinking.

The democracy of thinking manifests itself in L. Zadeh's approach, outside Aristotle's logic. This happens in cases when there is a chance (potence) to insert other choices (a third, fourth...) in-between the two poles. "Occam's razor" is thrown aside. This is enough to bring us to the threshold of the logic of tolerance. The reason why there is no doubt about the originally democratic nature of the artistic mind that gave birth to the text of "The Book of Dede Korkut" is that it is not always possible to exclude a third here. There is also a third between the two poles here, sometimes there is also a fourth and a fifth.

* * *

In this part we are going to discuss the exception of the law of the excluded third itself in the text of "The Book of Dede Korkut" from the perspective of fuzzy logic. In the gap between the two poles we are going to set forth in search of a third, fourth... contrary to Aristotle's logic.

So, we start a journey along the fuzzy paths to the ancient world of Dede Korkut leaning against the first logic facet of fuzzy logic.

Suppose there is a person standing in front of us. The man in front of us is either dead or alive according to our experience and Aristotle's claim. Yet we know that life is more colourful than these two poles. At least, this man can be neither dead nor alive, as stated in the text of "Dede Korkut". He might be in the state of "little death" or in the language of modern medicine, in "a coma". At this point, the state of a coma is a third state in the logic chain. Whereas Aristotle's logic excludes the third state (in our case—a coma).

The ideology *"either... or..."* has left us nothing but life and death. Contrary to this, fuzzy logic gives us the third. It informs us that there is also another reality where there is neither life nor death. This reality is in the middle of the two poles and that can manifest itself as the little death (a coma). The start of the journey with this scheme of "neither... nor..." (including a third) implies democratic sprouts in mind.

The comparison of the scheme "neither... nor..." with the scheme "either... or..." leads to interesting findings.

The scheme "neither... nor..." makes it possible to definitely accept a third betwixt. This is an indicator of its syntactic intention. "Neither dead nor alive" unambiguously entails the question "And what is it then?"

Whereas, there is no space for such a question in the scheme "either... or...". There the choice is certain from the onset. It is one of the two poles. That is why, one can say with determination that the scheme "neither... nor...", which is the frame of manifestation of fuzzy logic, generates an objective chance for a much wider choice. And as we have just stated, it implies a democratic mindware (democratic sprouts).

* * *

When we say that the elements of fuzzy logic in the text of "The Book of Dede Korkut" are not few, we mean that the ancient Oghuz used to have a need and passion to see the world, the relations among people more objectively, more adequately and more appropriate to reality. The ancient Oghuz do not reconcile with the logic consisting of the dual confrontation in the scheme "either... or...". They set forth in search of a third, even a fourth in the abyth between the two—"either... or...". There naturally appears a scheme "neither... nor...". The search leads them to this scheme. "Neither... nor..." reveals what is not allowed by "either... or..." and makes an attempt to fill that space in the middle.

Start with, we should note that "the search of a third' and its disclosure actually re-confirm the fact that the text we are dealing with belongs to the most ancient mythological prehistoric periods and is accepted as the reflection of "tracking" of what we see in the world by the democratic processes ongoing in our mind. Initially, this "tracking" is carried out poorly. Our mind is lagging behind from adequate understanding of the reality, and our language is behind from adequate description of our mind. Let us try to recall: Fuzuli, who said "Don't be deceived, the poet's word is certainly a lie", expressed an attitude towards this confrontation (language and mind). Fuzzy logic hurries to render an assistance to the bearer of the thought and language at this very moment.

We rely upon one of the basic principles of fuzzy logic—the principle excluding the law of the excluded third. As we wrote a bit earlier, the law of the excluded third suggests that no interim case is possible between the two contradicting poles. However, the formal logic (Aristotle's logic), which sees only a gap between "either... or...", is not always right. Actually, by this, fuzzy logic showed that "The excluded third" also has its own exception. Likewise, there is a little death (a coma) between death and life, if we proceed from the logic in the text of "The Book of Dede Korkut". Like the red tent which was put up between the white tent and the black one pitched at Bayandir Khan's gathering. Like Wild Dumrul's wife, who is ready to sacrifice herself for her husband, unlike his father and mother. Like the evidence presented by Dirse Khan's wife as the reasons for childlessness. The text of "The Book of Dede Korkut" generously bestows upon us such kind of examples.

Generally, final judgements should always be treated with doubt. They may have an unexpected "past". The Introduction of "The Book of Dede Korkut" has an abundant material regarding such final judgements. Among them, the ones which are in the Introduction and do not seem to allow any doubt at first sight are such kind of sentences:

> *Çıxan can geri gəlməz!*
>
> *Ölən adam dirilməz! [The man who dies is not brought to life, the soul which goes out does not come back...].*

Everything seems to be clear. It seems as if the recorded were a simple truth. Nevertheless, there is an issue not clear yet. Why do these statements address the Oghuz community?! If we consider Dede Korkut as one of the first teachers of this community, what lesson did he intend to teach through this statement?! Was there any need in such a statement in the early ancient period when the Oghuz community began its formation?! In fact, this statement is not without its own strange past.

Hence, if the judgement was uttered with such a determination (it is even possible to notice its acute intonation, though in writing), it means there was a need for such a statement. Having such a statement (information) was extremely needed by someone who was a member of the primitive society and encircled by mythological views the way Laocoon entrapped by snakes remained groaning. For a member of society like him/her, this statement is informative, it is new information. Since, he/she was convinced that the soul leaving the body (spirit) would return. Likewise, Dirse Khan's wife was convinced that the black mountain was lying asleep in the opposite side ("Qarşı yatan Qara dağdan bir oğul uçurdunsa, degil mana!" ['If you have hurled our only son down from yonder many-coloured mountain,... tell me.]) and that one day it would wake up. Likewise, this ancient man believes in the resurrection of the dead person [2].

This member of society should be brought to real life. Meanwhile, he/she is not in real life, but surrounded by metaphors, mythological thought (nets). The way of coming to real life was shown by Dede Korkut. He was teaching the member of of society a lesson on how to live in real life, away from dreams, mythological views. Was there any need for it? Yes, there was.

If someone gives us such information and says, "The dead man will not resurrect", it may seem strange to us and actually, this would not give us any information (new information). We have long (!) believed that the dead man will not resurrect, the soul coming out of the body will not return. However, for the ancient primitive period this piece of news had the content of being informative, and fresh. Thus, while the knowledge symptom of the ancient period was shaped in the pole *"the soul coming out will return"*, the symptom of the new period (the period of writing) is being shaped in the pole "*the soul coming out won't return*". And thawing of one of them within the other occurs by the principle of historism. And infinite existences in-between!! An infinite number of exceptions!!

The view *"The dead man will resurrect"*, which is an example of fuzzy logic and mythological view, turns into an example of formal logic over time ("*The dead man won't resurrect!")*. The final judgements in the text of "The Book of Dede Korkut" *("The soul coming out won't return!")* are the challenges to put an end to this view. As our mindware is developing, fuzzy logic sometimes leaves its place to the logic we are used to (formal logic) this way.

When speaking of the nature of women, the text of "The Book of Dede Korkut" says the following to the neighbour from the tongue of one of them in the Introduction part of the Epos. This woman is the one who leaves home in the morning and returns home in the evening:

Ölməyə-itməyə getməmişdim. Yatacaq yerim yenə bu xaraba olasıydı.

[I hadn't gone out to die and vanish for ever. I still have to sleep in this ruin].

Here is the outcome of the text. According to the internal logic of the Epos, if a woman is out of sight, she is either dead or got lost *("I hadn't gone out to die and vanish for ever.*"). Here one more (another) variant is provided by the discourse of the text (the subtexual meaning of the sentence it brings with itself) [3]. There is some scurry till "noon" between the two. And this sense enters as "a third" between the poles: *"To die—to scurry round till noon—to vanish!"* This interim variant brings the main function of fuzzy logic between the poles—the diversity in life. To be more precise, it brings a piece of the diversity in life.

Let us turn to another chapter (story). A bit earlier, above we already expressed an opinion. Following Bayindir Khan's gathering, Dirse Khan looks for the reason or who was to blame for his not having a son or a daughter and comes to such a formal conclusion: "*Bu qara eyib bana ya bəndəndir, ya xatundandır*". [*"This mysterious affliction of mine is due either to me or to my wife."*]

According to this logic, poles "*either me or my wife*" are shaped in Dirse Khan's mind. Dirse Khan's "clear" logic does not have any other reason, it cannot have any at all. An utter follower of Aristotle! The exception is (or exceptions are) introduced by Dirse Khan's wife. She actually acts as a representative L. Zadeh's logic and indicates several variants to Dirse Khan as a reason:

(1) Slaughter of horses the stallions, of camels the males, of sheep the rams, when you see the hungry, fill him;
(2) When you see the naked, clothe him;
(3) Save the debtor from his debt.

The reasons introduced by his wife appear as exceptions and thus, insert fuzzy logic or the exceptions just mentioned between the poles (*"me–my wife"*) regarding Dirse Khan's quest for the guilty (guilt). This way, there is an alienation from the preciseness *"either me or my wife"*, the colourlessness (*white–blackness*).

Another interesting example of the discovery of fuzzy logic is embodied again by the end of the same chapter. Dirse Khan has wounded his son Bughach who remains wounded in the hunting site. His mother sets forth to find him, and reaches the foot of Kazilik Mountain. In the text it says:

> *Dirsə xanın Xatunu... Qazlıq dağına gəldi, çıxdı. Alçaqdan uca yerlərə qalxıb çıxdı. [Dirse Khan's wife... reached Kazilik Mountain and she climbed it. Up she galloped from the lowlands to the high places].*

In our view, the name of another site, which is understood and accepted as "neither like this, nor like that", appears between the dual pole, i.e. between the lowland and mountain. In modern language this site is called *"a hill"* [*təpə*]. The matter is we encounter the word "a hill" in some other place in the Epos. That could have been used here as well, but it isn't. Therefore, we postulate that the expression *"from the lowlands to the high places"* is used instead of "*a slope*" [*yamac*] [4]. The text of "Dede Korkut" uses, thus, the expression "from the lowlands to the high places" instead of "a slope" [yamac]. The poles with exceptions are shaped as follows: **"a low site—a site higher than the low site—a mountain"**. In this mode the objective world gets its more adequate appraisal in the eye of man (the community where Dede Korkut lives).

Again in the same chapter the wounded Boghach answers his mother's question "Who has done it to you?" as follows,

> *Arslanla qaplanına qarğımagil,*
> *Qazlıq dağının suçu yoxdur.*
> *Qarğarsan babama qarğa -*
> *bu suç, bu günah babamdandır.*
> [*Curse not its lions and tigers;*
> *Kazilik Mountain is guiltless.*
> *If you must curse, curse my father.*
> *This crime, this sin is my father's*].

From the mother's point of view, either the lion or the tiger of Kazilik Mountain is guilty. There is a logic *"Either the lion or the tiger!"* in mind. In this case the mother proceeds from formal logic. Actually, there is some room for an exception here as well. That exception is brought to the chain of logic by Bughach: he mentions the guilty and says, "*my father*". The chain of fuzzy logic with the poles as well as an exception is shaped as follows: *"the lion–my father–the tiger*".

In another chapter when Salur Kazan's house is pillaged, he gives such an explanation to Karajuk the shepherd before the battle against the enemy,

> *Anamı düşməndən istəyəlim, savaş zamanı at ayağı altında qalmasın [I shall ask the infidel for my mother ((so that she is not trampled under foot by horses))].*

Salur Kazan promises the infidels that he will return (actually, this is a military trick [5]) after taking his mother from the enemy, and he claims his mother back. Whereas, the infidels react to Kazan's desire as follows,

> *Qarıcıq anani gətirmişiz – bizimdir. Sana verməyiz. Yayxat keşiş oğluna verəriz.... [Your little old mother we have taken; And she is ours. We shall not give her to you. We shall give her to the son of Yaykhat the Priest...]*

Clear logic says,

> *Ya sənə verəriz, ya bizə verəriz (bizdə qalar).*
>
> *[Either we shall give her to you, or we shall give her to us (we'll keep her)].*

Immediately the expression "we" is divided into*"you–we—the son of Yaykhat the Priest"* and more precise contours of the reality, but more fuzzy contours for the chain come to the surface. The principle of a kaleidoscope starts functioning. A hairy and ugly face of the third component (the son of Yaykhat the Priest) appears out of the diffused "we".

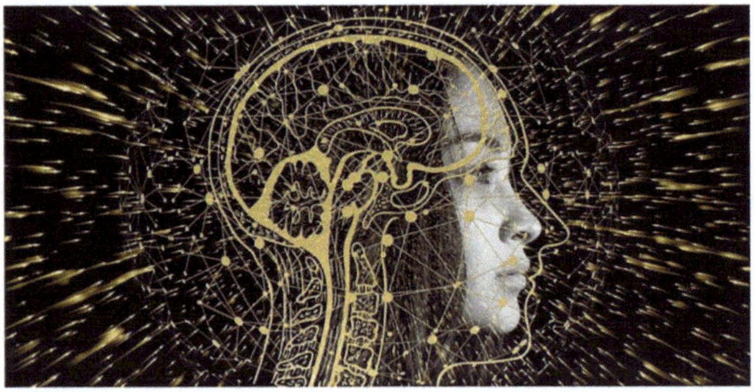

In the logical chain the poles take vividly the exception between themselves in the chapter "Wild Dumrul". A bit earlier, we mentioned about it. Allah, whose anger to Wild Dumrul subsides, commands Azrael, "*Dəli Domrul canı yerinə can bulsun, onun canını bağışladım"*. [*"Let him find a soul in place of his own, and his own soul can go free."*]

As soon as Wild Dumrul learns about this command of Allah, such a chain of formal logics is embodied in Wild Dumrul's mind. He says to Azrael,

Məgər bir qoca babam, bir qarı anam var; gəl gedəlim, ikisindən biri bolay ki, canını verə, alğıl, mənim canımı qoğıl. [But I do have a father and mother; come, let's go; one of them might give his soul, then you take it and leave mine].

The logical chain is *"either father or mother*". Then an incident like a flash of lightning happens between these poles. "The third" moves itself forward. As an exception, it gets an immediate and non-hesitant place in the text. This exception turns out to be an unimagined person, Wild Dumrul's wife who is considered a girl of not their tribe. Unlike Wild Dumrul's father and mother,

she is ready to sacrifice her life in place of Wild Dumrul to Azrael, saying, "Let my life be sacrificed for yours." Thus, the exception of the excluded third (i.e. "there is a third!") finds its reflection in this example extremely vividly which is the very backbone of fuzzy logic.

We often encounter the exception of the law of the excluded third all through the text of "Dede Korkut". As we mentioned at the beginning of this chapter, this should be appraised as an attempt of the author (authors) of the text of "The Book of Dede Korkut" through time to see the world, the surroundings more diversed, as a loyalty to fuzzy mindware. This attempt can be accepted as a sign hinting at the presence of democratic shades in the depth of our mind since the ancient times. The study of these and many other interesting points in fuzzy logic opens new horizons before the study of "The Book of Dede Korkut", linguistics, as well as artificial intelligence and information theory.

References

1. Zadeh, L.A.: Is there a need for fuzzy logic? Inform. Sci. **178**(13), 2751–2779 (2008)
2. In the Italian city of Cerveteri the ancient inhabitants of the city, Etruscans, have a subterranean town. Now it is a museum. Walking along the streets of the city, you see houses, you can enter any of them you like and in one of the rooms there you'll see the dead family member's carcass (the carcasses) which is all bones. This is the expression of belief implying that the dead person will resurrect one day. Such state is observed in ancient, primitive societies
3. In our opinion, discourse is not a repetition of the textual elements this way or another. Discourse is a basis of additional knowledge the text brings (most often secretly) with itself
4. It was Prof. Dr. Ahmet Bican Ercilasun who focused our attention to the word "yamac"
5. Abdulla, K.: The Semantic Twinkling in "The Book of Dede Korkut". Mytholinguistics, p. 164. Baku (2022)

Chapter 7
“My Beauty, You Are More Beautıful than the Beauty of the Beautıes!”

7.1 Fuzzy Sets in “Dede Korkut”

They say, one day when speaking of his wife Fay Khanum, Lotfi Zadeh, the founder of fuzzy logic, outstanding scientist, praised her as follows,

> O, gözəllər gözəlindən də gözəldir. [She is more beautiful than the beauty of the beauties].

Most probably without even realizing it, the great scientist used such a metaphor that it would be impossible to think of any other attribute higher than that to express this sense. What else can be said about a beautiful woman? Perhaps nothing. It is absolutely impossible to surpass the level of being more beautiful than the beauty of the beauties. One of the principles that stand in the foundations of fuzzy logic (the fuzzy set facet) is formed this very way.

It is known that the theory of sets developed by the outstandıng mathematician G.Cantor constitutes the basics of the model of the mathematic world which is not that much real and gives a quantitative description of the real world. It can be simplified and expressed approximately as follows, “If we subtract an infinite set from the infinite set, the remaining will be again an infinite set.” [1] A nice formula to understand infinity! Let us recall Einstein’s idea which we have quoted realier.

Why is the mathematical world built on the basis of sets not real, why is it far from our real world? For, classic sets have precise boundaries, cover only quantity. For instance, if out of the universum of entire women in the world, we appraise an element, i.e. a woman according to outward beauty, in G. Cantor’s notion of a set, she is either beautiful or ugly. In other words, she is either within the set of “beauties”, or outside it.

Whereas, a mathematical world based on the human-oriented mathematics, which leans against fuzzy set, is closer, more adequate to the present-day world we live in. Why? Since, in this world expressed with symbols, texts, words like in real life, quantity is in unity with quality. In other words, the main message of the fuzzy set facet of fuzzy logic is that everything, every phenomenon in the world except Tengri

K. Abdulla and R. A. Aliev, *The Book of Dede Korkut and Fuzzy Logic*, Studies in Fuzziness and Soft Computing 438, https://doi.org/10.1007/978-3-032-05906-2_7

has a degree. If we view a fuzzy set, all the elements in the universum have some degree of identity to this set. If we return to the previous example, that helpless ugly woman was perhaps not very ugly, with her certain degree (degrees) she could as well belong to the set of the beauties (let us recall the expression "Judge Leyli with Majnun's eye").

The key principle of the fuzzy set facet of fuzzy logic induces us to strive for the highest level in the gradation of sets and find a language unit appropriate to it. For this purpose, one should define the initial and final poles. In our example above, these are the poles "*beauty—the beauty of the beauties*". It is at this time that the search of fuzzy logic starts. How can one leave this "beauty of the beauties" behind and approach the notion (and the reality) "much more beautiful"? Certainly, with L. Zadeh's expression "You are more beautiful than the beauty of the beauties".

* * *

Fuzzy sets are a reliable means encompassing the restrictions imposed on the linguistic information in the analysis of literary texts. It should be taken into account that when some incomplete information is provided regarding some action or phenomenon in the text, this decreases the trust in it (the text). For instance, the expression "certainly/evidently" is semantically stronger than the expression "probably". At this point, one should take into account also such determiners as "*little*", "*much*", "*more*". It should also be noted that actually, the author provides information in the text on the basis of his/her epistemological position based on his/her knowledge, observation or belief.

And now let us set forth from the second facet of fuzzy logic—the fuzzy set facet towards the world of "Dede Korkut".

It is sometimes easy and sometimes difficult to define the degrees between an ordinary state and maximal state in the text of "Dede Korkut". For instance, suppose there is a goal to generate a vision of a big, very big ship through some phrase. The language capacity of the text allows us to describe it like this. The text of "The Book of Dede Korkut" would easily express this image in the vision as follows, ***"Böyük ağac gəmilər oynadan su"*****.** **[((*"Water in which big wooden ships play."*))]**

However, the ship in the vision is not merely big. It is unconvincingly big. The language in the frame of the text attempts to adjust to the image in our vision. In order to introduce immensity, it repeats the adjective. Therefore, we read in the text,

"***Böyük-böyük*** *ağac gəmilər oynadan su"* [*Water in which ((big and big)) wooden ships play"))*].

The expression "big and big" (***"böyük-böyük"***) is an effort here to imagine and express the enormous size of bigness. This is where fuzzy logic starts. Thus, the path leading to inifinity should pass through the regime of fuzziness. The reduplicative words in the language come to a complete rescue here in order to realize this case.

Sometimes it so happens that some words enter between the poles (as a third) and turn into the names of new notions. Distinguishing the "low" and "high" places enables the appearance (!) of the name of a new topological site. We would like once more to refer to the point above we dealt with a bit earlier. Dirse Khan's wife is looking for her wounded son Bughach on a low site (plain) at the foot of Kazilik Mountain, but fails to find him. What does she do? To see a wider area she ascends from a lowland (plain) to a high place. This is such a place that you can climb galloping on a horseback too. In the text it says,

> Xatun alçaqdan uca yerlərə çapıb çıxdı [Up she galloped, from the lowlands to the high places].

Such kind of place can only be "**a hill**". Thus, between *"a lowland"* and *"a mountain"* there appears to be a place higher than the lowland. **"A low place—a place higher than the low one—a mountain**" is actually an order of some notions in the system of sets and the insertion of a third place in-between the two poles (a *place higher than the low one,* i.e. *"a hill"* [təpə]) results in the division of this order by fuzziness into a new degree.

Here it calls into question. As the word *"təpə"* [a hill] passes through the text of the Epos, the expression *"a place higher than the low one"* can be used as a replacement for the word "*a slope*" [yamac] instead of "*a hill*" ["təpə"]. In that case the logical chain is built in the following way: "*düzən—alçaqdan üca yer (yamac)—təpə—dağ* [*a lowland—a place higher than the low one (a slope)—a hill—a mountain"*]. Deviation from the exception (already with the participation of a third, fourth) presents itself brightly again.

* * *

In the text of "Dede Korkut" there are such lines associated with the praise of Allah at different points:

Ucalardan ucasan,
Kimsə bilməz necəsən,
görklü tanrı!
Neçə cahillər səni
Gögdə arar, yerdə istər –
Sən xud möminlər könlündəsən.
[You are higher than the high,
No one knows what You are like,
Beautiful God!

Many the ignorant who look for You in the sky or seek You on earth,
But You are in the heart of the Faithful].

"Uca" [High] is already a pole. Beyond it the language capacity does not allow to define a new degree through a new word. Therefore, being deformed, the word "high" denotes the highest superlative degree. And here the expression **"Ucalardan ucasan" ["You are higher than the high"]** appears as the highest degree of superiority. It is curious that in the second part of this example we encounter the deviation from the law of "the excluded third" of fuzzy logic, i.e. the old facet of fuzzy logic familiar to us. Allah is neither in the sky nor on earth. Once Allah is in neither of these poles, where is Allah then?

Since, according to Aristotle's logic the third is excluded and Allah can be nowhere else then. Allah can be in either one of these two places. No, it turns out that there is also a third place. Allah can also be **in the heart of the Faithful,** as it is claimed in the Epos. This third site is the very unexcepted "third" included into the chain of logic.

In the Epos there is a description of **a lamentating situation** ["*şivən*"] in association with different moments. In the Epos this word is used in its ordinary form as "**lamentation**", (for instance, *"evinə şivən girdi"* [*"lamentation found its way to his home"*]). As the highest degree of this word, the notion *"big lamentation*" appears in the text as an expression "*black lamentation*" [2]. Regarding this, we seem to encounter the highest degree of fuzzy logic. The poles *"lamentation–black lamentation"* are formed by this logic. There is no third. Not yet. In the text of "Dede Korkut" *"black lamentation"* itself turns into an interim state, stops being a pole. What happens to the other pole then? Is there any degree stronger than that? It turns out, yes, there is.

In the text it says in the chapter "Wild Dumrul" that by God's command a fine handsome warrior dies by the slope of the bridge built over a dried-up stream by Wild Dumrul. In his lifetime everyone would love that warrior. Therefore, *"ol yigit üzərinə möhkəm qara şivən oldı"* [*Great was the black lamentation over that warrior*].

Neither "*lamentation"*, nor "*black lamentation*", but the very *"great black lamentation!"* This is how the other strong pole is found. When saying "*great black lamentation*", there appears an expression corresponding to a higher degree of "**lamentation**", probably following the highest gradation of language capacity. Taking the mid position, **"black lamentation"** ceases being a pole, carries out the function of an exception, leaving its previous place (of being a pole) to "**great black lamentation**".

In the text the highest degree sometimes appears not through a particular word or expression, but through a syntactic-semantic order. In the Epos when the nobles of the Oghuz go to help one another, they were quite big in number. It would not be possible to finish counting them, i.e. to count and say a certain number. Language capacity can provide such a variant corresponding to this situation (point), ***"Sayılmaqla Oğuz bəyləri tükənməz"***. [*"Of counting the nobles of the teeming Oghuz will not end."*]

This is a firm decision. Yet, there is a milder, in other words, future-oriented syntactic variant too. In the text it says, putting aside the firm decision: "**Sayılmaqla Oğuz bəyləri tükənsə olmaz**". [*"Of counting the nobles of the Oghuz there could be no end."*] The difference between *"will not end"* and *"there could be no end."* in the two sentences is interesting.

Between the two alternatives the difference in favour of what is in the text is nourished from the idea of infinity to be brought by "olar" [possible/could be] hidden within the potential of the word "olmaz" [impossible/could be no].

The main detection of fuzzy logic takes place through the participation and assistance of some language elements. The detecting element appropriate to this chapter is reduplicative words. In language reduplicative words act as a means increasing the intensiveness of the sentence element they belong to. Owing to them the expressiveness of the text grows, its literary-stylistic "corpus" consolidates. However, prior to any stylistical nature, it should be noted that at the primary moment, owing to reduplicative words, the inclination of our consciousness towards fuzziness is also expressed.

The text of "The Book of Dede Korkut" is also abundant in reduplicative words. Their usage is aimed at pointing out "the work" of the language component to which they belong in its full capacity. The reduplicative words used here serve rather to the disclosure of fuzziness. For instance, let us consider these examples all through the epos "The Book of Dede Korkut" [3].

Quru-quru çaylara su saldım [((I brought water to the rivers that had run **dry and dry**))];

qanlı-qanlı sular [((**bloody and bloody** waters))];

sən gedəli **gumbur-gumbur** davullar döyülmədi [since you left the **thundering** ((**thunderous and thunderous**)) drums have not been beaten];

soyuq-soyuq sular [**cold and cold** stream];

taşqın-taşqın sular [((the waters that run **flooding and flooding**))];

açuq-açuq meydana bənzər sənin alıncığun [your dear forehead is like a **broad open** ((**open and open**)) field];

yüksək-yüksək qara dağlar [black mountains **peak on peak**];

böyük-böyük suların köprüsü ağac [wooden the bridge over **mighty** ((**big and big**)) rivers];

qara-qara dənizlərin gəmisi ağac [wooden the ships on the **black** seas// ((**great and great seas**))];

yumru-yumru ağladı [wept **bitter** tears//((shed **big round and round** tears))];

çığnam-çığnam qayalardan çıxan su [((the water originating from the rocks **bubbling and bubbling**))];

içib-içib Ulaş oğlu Salur Qazanın alnına şərabın itisi çıxdı; [Salur Kazan son of Ulash drank **deep/ ((drank and drank))**, and the strong wine rose up to his forehead.]

yata-yata yanımız ağrıdı, **dura-dura** belimiz qurudu [Our sides are sore with **too much reclining/((reclining and reclining))**, our backs are stiff with **too much standing/((standing and standing))**.];

bir gecə döşəyində **qatı-qatı** inlədi [one night he cried **hard/** ((groaned **deep and deep**)) in his bed];

qalxıbanı iki qardaş **quca-quca** görüşdülər [((the two brothers got up and **embraced**/met **hugging and hugging** each other))];

bu məhəldə Oğuz ərənləri **alay-alay** gəldilər [Then the Oghuz warriors came, **regiment after regiment**...]

Through reduplicative words (actually, through the mechanism of repetition in it) the ordinary state is consolidated, i.e. is exalted to its highest degree. Salur Kazan the son of Ulash drinks not just simply, he drinks again and again, he drinks continuously. The Oghuz warriors arrive not in one regiment, they arrive in numerous regiments. All the adjectives, nouns, imitative words (**bıldır-bıldır [((drop by drop))], zar-zar [((sobbing and sobbing))], ökür-ökür [((weeping and weeping**))] denote the high, boundless boundary of the description through their own repetition (by forming reduplicative words). Repetition manifests itself as a reliable "construction material" of fuzzy logic.

There is sometimes no need left even for repetition. At the point where fuzzy logic should "work" (is expected to work), an odd variant appears instead of repetition. The weight of the preceding repetitions has an impact also on the element in the last!!! link. When reaching the point of the sheep to be sacrificed after "**Qatar-qatar** dəvələr, **tavla-tavla** şahbaz atlar, **soyuq-soyuq** pinarlar" [**caravan on caravan** of camels, falcon-swift horses, **stall after stall** of them, **cold cold** streams ((springs))], the expression "*ağayılda ağca qoyunum*" [*"my white sheep in the folds"*] is used. This poor sheep is actually not alone. They are also *pens after pens* [*tövlə-tövlədir]*. Nevertheless, even in its singular number, the sheep can generate intensiveness, the highest degree, fuzziness (unaccountability) in the light of the preceding sacrifices.

An attempt to reach the highest boundary, in our opinion, stops the neutral flow of thinking, bringing it to the edge of the highest "boiling" degree. This can be appraised also as deviation of the internal energy of the text from an ordinary state.

As Molla Nasraddin expresses it, when his wife asked Molla Nasraddin why he did not get out of the circle drawn by the outlaws who looted them, he answers, "Three times I took my foot out of the circle and thus, went against their "directives"—(orders)." Molla Nasraddin must be one of the first characters among all the characters we have met in literature who substantiated his actions with fuzzy logic. Molla Nasraddin, who puts his foot out of the circle drawn around him three times, is a man of "courage" who attempts to introduce some degree between the poles "**çıxarma-çıxar**" ["do not put out–put out"].

The principle of fuzzy logic in this facet leads to the formation of "simple–complex" relations ongoing in the ancient man's mind, and polishes different variations of the transition from simple to complex. In other words, it forms the hidden potential of this. It also presents how the processes ongoing in our mind are reflected in our language. The prevention of the neutral simple flow of mind (its complication by inserting a degree in-between the poles) will result in the establishment of style, formation of language elements, as a whole enrichment of our language in the subsequent stage of language development. As for now, as it is obvious, in the competition of "language–mind", language is vividly lagging behind mind [4].

So, in the light of what we have said, let us note by returning to the great scientist Zade's statement "My beauty, you are more beautiful than the beauty of the beauties", even the expression "the beauty of the beauties" is not the last boundary of that vision, the existence of a degree higher than that is confirmed by the language material. This is the expression "the beauty of the beauties". Even the expression "more beautiful than the beauty of the beauties" seems not to remain as the last one in this gradation. And the great scientist, great personality, great lover L. Zadeh was right to select for his wife the expression of the highest degree existing in language so far.

References

1. It is not an easy matter to understand and accept an infinite multiset. This law of Cantor was once of great importance for one of the co-authors of this book (Kamal Abdulla) to understand and accept infinity. See source #7, p. 32
2. In Old Turkic languages the word "qara" expresses also the meaning of "big". For instance, the word "qara" in *"Qara dəniz"*, *"Qara hisar"*, *"Qarabağ"*, *"Qaraqum"* implies "immensity" apart from denoting the colour
3. The meaning expressed by reduplicative words in the examples is sometimes conveyed through other means rather than reduplication in Geoffrey Lewis' translation
4. Chomsky, N.: Language and Mind, pp. X, XI. Cambridge University Press, Cambridge

Chapter 8
"The Book of Dede Korkut" in the Epıstemological Light of Fuzzy Logic

8.1 Fuzzy Epistemology

In this book we set forth to the world of "Dede Korkut" through the general introduction of fuzzy logic introduced in the previous chapters and in the light of its two facets—the facet of logic and the facet of fuzzy sets. In this chapter we'll start searching for the illimitable information in the texts of "Dede Korkut" and the meaning in their substrata, proceeding from the third facet of fuzzy logic—the epistemic facet.

The epistemic facet of fuzzy logic is associated with the description of the knowledge presented in the texts, the semantics of natural language, and the analysis of the information in the texts. In this aspect, natural language is accepted as a system describing the perception information, i.e. the information that cannot be measured directly. This facet is founded not on the theory of probability, but the theory of possibility. When speaking of the epistemic approach, our main goal will be to bring out of the concealed inside of the text the information the author was reluctant to say or intended to hide, and appraise it.

Fuzzy logic fills in the gap between the real systems observed at the expense of scientific assumptions and their phenomena (the poles). In the history of philosophy, science and technology these gaps enjoy great potential for epistemological discussions.

Anything epistemological does not mean understanding it absolutely as "an object" existing in a narrow sense, but stipulating its imagined image, and functions [1]. Since it is not possible to clarify it from the beginning, this represents itself initially in such a feature and uncertainty that it makes human searches unquestionable.

K. Abdulla and R. A. Aliev, *The Book of Dede Korkut and Fuzzy Logic*, Studies in Fuzziness and Soft Computing 438, https://doi.org/10.1007/978-3-032-05906-2_8

That something epistemological is of a fragile status does not mean that it remains only as concealed and can be revealed completely only through perfect manipulations. As an object, it does not exist as a readymade, but can exist only in consciousness as a conceptual side of the object (like an image in the brain).

Fuzzy epistemology helps to define (to draw away) the complexity, uncertainty (the fog) in the text.

According to fuzzy epistemology, both uncertainty and partial truth are needed to understand our attitude towards indefinite provisions. For instance, if "a tall man" has a classic height, then there is also a tall man with the shortest height. Certainly, to find out how high the shortest man is, we have no solution. One of the co-authors of this book had once published a story under the same title: "The Tallest Dwarf in the World" [2]. Back then the comparison of the idea of the height of shortness with the image in mind seemed intuitively curious.

In general, the demarcation of some border between the tall and not the tall can be random. Also, no concept associated with our usage of the word "high" ("tall") is able to draw to attention the special point of boundaries. Thus, we can say that the fuzzy epistemological field encompasses the semantic boundaries whose semantic limits are unknown and which seem random in texts.

8.2 Epistomologic Facet of Fuzzy Logic in "Dede Korkut"

Let us consider an example from the folklore:

Oğlan, qaşımı neylərsən, oğlan?
Göylərdə hilalı görməmisənmi?
Oğlan, ləbimi neylərsən, oğlan?
Məgər beçə balı görməmisənmi?
[Oh boy, why are you interested in my eyebrows?
Haven't you seen the half-moon in the skies?
Oh boy, why are you interested in my lips?
Haven't you tried the honey of lady honey bees?].

Based on this example, the epistemological presentation of fuzzy logic focuses the attention on the state that the girl reciting the poem is concealing her eyebrows and says to the boy that instead of her eyebrows it is sufficient for the boy to see the half-moon in the sky, "Once you have seen the Moon in the sky, you can imagine my eyebrows too." However, the object is not present, she is invisible. To be more precise, she is not described quite adequately. The same situation is also repeated regarding the girl's lips. The lips are resembled to the young bee's honey. The direct description of the lips is concealed by the author and by imagining the lady honey bee's honey, fuzzy knowledge about lips is offered indirectly.

This description can be acknowledged as fuzzy logic constructed through the epistemological approach. The author tries to conceal the girl's eyebrows and lips, yet, at the same time, allows us to imagine them in a fuzzy way—the Moon in the skies and the honey of the lady honey bees start a claim of perfection about the imagined lips and eyebrows in this description and there appears a fuzzy vision close to the degree of perfection (by generating a different degree of perfection in various individuals).

The example under analysis reconfirms that with its various manifestations, fuzzy logic lives in the depth, the concealed stratum of the largest literary spheres.

The brightest example associated with the third epistemic facet of fuzzy logic in the text of "The Book of Dede Korkut" is the situation encountered by Salur Kazan and his warriors during hunting. To introduce to his son Uruz the sites where he had displayed knighthood, he goes out hunting with a small group of people, taking also his son with him. The infidel's spy espies them, like always before, and Salur Kazan is assaulted by an enemy of sixteen thousand troops. In the text it says,

On altı bin qara donlu kafər ata mindi. Qazanın üzərinə alqar yetdi. Baxdılar gördülər (baxıb görənlər artıq Salur Qazangildir), **altı bölük toz endi. Kimi aydır: keyik tozudur. Kimi aydır: yağı tozudur. Qazan aydır: Keyik olsa bir, ya iki bölük olurdu. Bu gələn bilmiş olun yağıdır.**

Sixteen thousand black-mailed infidels mounted their horses and charged down on Kazan. He and his men suddenly noticed that six separate clouds of dust had drawn near to them. Some said, 'It's the dust of a herd of deer'; others said, 'It is the dust of the enemy.' Kazan said, ' If it were deer there would be one or two clouds at most; you can be sure it is the enemy that comes.

Kazan does not see the enemy, dust is everywhere. Based on his preivous experience, from the signs of the dust—the clouds of dust being more than two, Salur Kazan knows that it belongs to the enemy. In a fuzzy way, Kazan "estimates" the enemy, an image he has not seen, but vizualizes.

It so happens that the information the author tries "to conceal" is revealed by one of the infidels.

While Salur Kazan is hunting, Shöklü Malik "brushes", as is described in the text, loots and takes away his property, money, his son, wife, mother, horses, camels. After wining and dining Shöklü Malik estimates his trophy and says,

Bəylər, Qazanın tölə-tölə şahbaz atlarını minmişiz, altun–axçasını yağmalamışız, qırx yigitlən oğlı Uruzı tutsaq etmişüz, qatar-qatar dəvələrini yetmişüz, qırx incə belli qızlan Qazanın halalını tutmuşız. Bu heyfləri biz Qazana etmişüz.

[Nobles! We have mounted Kazan's falcon-swift horses, stall after stall of them, we have looted his gold and silver, we have captured his son Uruz and his forty young men, we have led away his camels, file upon file, we have taken Kazan's wife and her forty slender-waisted maidens; these losses we have inflicted on Kazan.]

When Shöklü Malik said that, one of the infidels intervened. This infidel confuses the precise number of the trophy listed with precise logic. In the text it says,

Kafərin biri aydır:

– Qazan bəydə bir heyfimiz qaldı.

Şöklü Məlik aydır:

– Mərə aznavur, nə heyfimiz qaldı?

Kafər aydır:

– Qazanın qapulı Dərbənddə on min qoyunu vardır. Şol qoyunları dəxi gətirsək, Qazana ulu heyf edərdik – dedi.

[One of the infidels said, 'We have one loss yet to inflict on Prince Kazan.' 'Noble lord, what loss is that?' asked King Shöklü. 'Kazan has ten thousand sheep at the Gated Black Pass [Derband]; if we were to take them too, we would inflict great loss on Kazan.]

The difference between "*Heyf*" ["*inflicted loss*"] and "*ulu heyf*" ["*inflicted great loss*"] is measured by ten thousand sheep. The introduction of the number with such preciseness may seem too strange, but it is the brightest example of fuzziness. Although all the other trophies, even the number of horses and camels are not directly indicated, this does not attract the attention. *"Horses, stall after stall of them, gold and silver, camels, file upon file"* are already in the infidel's hands and he can specify their number. Whereas, ten thousand sheep are far away, not taken hold of yet, probably will not be possessed at all, actually they won't. Yet this number—the number "ten thousand"—is actually fuzzy, and this fact is defined through the epistemolgical approach.

* * *

Another point leads us to the chapter "Emren the Son of Begil". Here the difference between the image you construct in your mind and its replacement in life is described through the epistemological appearance of fuzziness.

Begil who guards the borders of the Oghuz falls from his horse and lies sick in bed. The spy of the infidels (*"Yarımasın, yarcımasın" ["May he never succeed in life!"]*) "is again at work" and informs them about Begil's sickness.

The enemy who would never dare to approach the Oghuz borders for fear of Begil gets encouraged and assaults Begil's land. However, Begil too has a spy of his own in the land of the infidels, and this spy sends a word to Begil that he should be careful. 'An enemy is about to attack you.'

Begil, who lies sick in bed, gives to his son Emren his armour to wear and his horse to ride. He disguises his son to look like him and sends him onto the enemy in his own stead. Emren goes against the enemy. Again, the dust would rise up to heaven like in the situation related to Salur Kazan. Emren becomes invisible from far away, i.e. he becomes unrecognizable.

Since, ***"haçan ki al ayğır (Bəkilin atı) yağı qoxusun alsa, ayağın yerə döyərdi, tozı göyə çıxardı"* [*"Whenever the red stallion [Begil's horse] smelled the enemy, he would paw the ground, and the dust would rise up to heaven*".]**

Again the horse remains loyal to his habit and paws the ground. Everything around was embraced in dust.

In the text it says,

Kafərlər aydır: Bu at Bəkilindir, biz qaçırız. Təkur aydır: Mərə, onat görün, bu gələn Bəkilsə, sizdən öndən mən qaçaram – dedi. [The infidels said, 'That horse is Begil's; we are off. Their king replied, 'Look carefully. If this man who is approaching us is Begil, I shall run away before you do'].

When it is impossible to identify with an ordinary eye whether it was Begil, the scout starts his mission. He scrutinizes and sees that although the horse is Begil's, but the man on it is not Begil. This way, the message reaches the king.

As Begil is in the king's vision, he fails to distinguish between Begil and Begil's son who was made to resemble him. Certainly, here the dust rising due to Begil's horse's pawing the ground also plays its part. All hopes are pinned on the scout. He states the truth. Further, the surroundings, especially the horseman ahead get their real, adequate dimensions. A piece of incomplete knowledge is removed for King Shöklü. The epistemological approach blows the misunderstanding into the air like dust. Begil is in Begil's place—in his bed at home. And Begil's son Emren gets his real features before the enemy. However, the space where Begil is really in is revealed by his son Emren to the enemy through another lie. In the text it says,

> **... Bəylərbəyi Salur Qazan, qardaşı Qaragünə, Dönə bilməz Dülək Uran, Dözən oğlu alp Rüstəm, boz ayğırlı Beyrək Bəkilin evində içərlərdi. Səndən casus gəldi. [...Salur Kazan, Commander of Commanders, Salur Kazan's brother Kara Güne, Dölək Evren who knows not retreat, Alp Rüstem, the son of Düzen, Beyrek of the grey horse were drinking in Prince Begil's house when a spy on you arrived].**

What is surmised from Emren's words, being busy with his guests, Begil had sent him—young Emren against the enemy carelessly. Begil's image is again lost in the midst of dust. Such kind of fuzziness—this is who Begil is, here groaning hard in his bed, and Begil is entertaining his guests—all these state the author's intention "to conceal" Begil from the vision of the enemy, come what will. Begil and his son Emren continue to replace each other like the twinkling stars. The schemes *"It is either Begil or Emren"* and *"It is neither Begil nor Emren"* replace each other as fast as lightning.

In the text of "The Book of Dede Korkut" the examples revealed with the help of the epistemic facet of fuzzy logic also show that in the depth of the ancient Oghuz' mind, the bounderies of reality and unreality are not utterly distinct from each other, they can transfer swiftly into each other. Even if the author uses such means as the horse's pawing the ground (the dust covers everything!), finally, a scout is needed to put everything in order and announce Begil's absence. What would have happened, if there had not been a scout?! Would the infidels have run off with King Shökli at their head, if they had mistaken Begil's son for Begil? This already implies the elucidation of quite another theme from the epistemological angle.

References

1. Abbasov, A: The Discipline of Lotfi Zadeh and Rafik Aliev—The Treasure of Knowledge Generating a New Civilization. İctimailider. Az. (2021)
2. Abdulla, K.: The Selected Works (in 7 volumes), vol. 2, p. 480. Baku (2020)

Chapter 9
The Relational Facet of Fuzzy Logic and "The Book of Dede Korkut"

9.1 "If...Then..." Rules

Fuzzy relation in "Dede Korkut"

The manifestation of fuzzy logic in a literary text is the introduction of the presented message mainly not in a concrete, but quite abstract (unfathomed) way. This is by principle the key line of the fourth facet of relations of fuzzy logic. This case is achieved through the known fossilized (but sometimes cracked) scheme "if... then...".

In this chapter we are going to speak about the manifestation of the derivatives of the fourth facet of fuzzy logic in the epos "The Book of Dede Korkut". Being the facet of fuzzy relations, the fourth facet of fuzzy logic embraces the fuzzy relations, more precisely, fuzzy interdependencies between the cause and effect. The foundations of this facet are fuzzy expressions (variables) and the relations based on the scheme "If... then..."

When speaking about the facet of fuzzy relations, we imply the close interconnection of some elements of the sets reviewed in different situations with one another this way or another. For instance, the relation between the set **"young people"** and "**experienced people**" can be:

> Əgər Salur Qazan oğlu Uruz çox gəncdirsə, o zaman o, az təcrübəlidir. Əgər Salur Qazan oğlu Uruz orta dərəcəli gəncdirsə, o zaman o bir az çox təcrübəlidir. [If Uruz the son of Salur Kazan is young, then he is less experienced. If Uruz the son of Salur Kazan is averagely young, then he is experienced a bit more, etc.].

Fuzzy relations are a very important conception in fuzzy theory. These relations realized with the scheme "if... then..." are extensively used in the study of real life processes in many fields.

K. Abdulla and R. A. Aliev, *The Book of Dede Korkut and Fuzzy Logic*, Studies in Fuzziness and Soft Computing 438, https://doi.org/10.1007/978-3-032-05906-2_9

It should be pointed out that we often encounter the mode of expression formed on the basis of *“If... then...”* scheme and transformed into a sentence structure in the epos “The Book of Dede Korkut”. There are even such structures which have acquired a form astranged from the initial scheme. As if the scheme were “cracked”. Nevertheless, it is not hard “to recognize” them, i.e. these relations. Let us consider such an example:

> **Ol zamanlar oğullar ata sözünü iki eləməzdi. Eləsəydi, ol oğulun üzünə baxmazdılar.**
> [In those days a son did not go counter to his father’s words. Had one done so, the father would not have acknowledged him as his son.]

There is no doubt that the scheme ***“if..., then...”*** stands in the substructure of the mindware which generated this sentence. Although the sentence seemed to have been exposed to “a syntactic harakiri”, it doesn’t cause any challenges to return the sentence to its initial substructure. For this purpose, we’ll need to expose it to some “operation”. Following “the operation”, the sentence would acquire such a state:

“***Əgər ol zamanlar oğullar ata sözünü iki eləsəydi, o zaman ol oğulun üzünə baxmazlardı***”. [“*If in those days a son went counter to his father’s words, then the father would not have acknowledged him as his son.*”] This substructural variant (actually, the invariant) is easier, more favourable, whereas there is no doubt that the latest variant has been through some literary embelishment. Let us consider some other examples:

"Ulaşıban sular daşsa, dəniz dolmaz;

Yapa-yapa qarlar yağsa, yaza qalmaz;

Qaravaşa don geyirsən, qadın olmaz" və s. və i.a.

[Though the rivers rage and overflow, the sea is not filled.

Though the snow falls in huge flakes it does not last till summer;

Though you dress a captive girl in a robe she does not become a lady, etc.].

All these examples can be fitted into the main scheme of the fourth facet of fuzzy logic. Then we are going to encounter such a scene:

Əgər ulaşıban sular daşsa, o zaman dəniz dolmaz;

Əgər yapa-yapa qarlar yağsa, o zaman yaza qalmaz;

Əgər qaravaşa don geyirsən, o zaman qadın olmaz.

[If the rivers rage and overflow, then the sea is not filled.

If the snow falls in huge flakes, then it does not last till summer;

If you dress a captive girl in a robe, then she does not become a lady, etc.].

* * *

One might wonder: anyway, in the field of linguistics we learnt the sentences formed on the basis of this scheme. We pointed them out as conditional or tense clauses. So, what changes?

Yes, this is indeed the case. Yet, what changes is the fact that so far linguists have not appraised this type of sentence proceeding from the ideology of fuzzy logic, they were far beyond such an ideology. These sentences were not considered as a result of some processes ongoing in some depth of mind. Now even the part of the circle which has remained open is finally being closed, there appears a systematicity, wholeness. The matter is already not a conditional or tense clause, but the democratic processes ongoing in our ancient ancestor's mind. For now, this is the answer to the question "What changes?"

Being not categorical, the fourth facet, as well as the other facets of fuzzy logic are closer to the contemporary man. It is like this that their ideological status can be characterized. Being away from categoricalness implies being away from determinism, dogmaticism, totalitarianism, and finally, if we transfer it to the description of psychological states, from selfishness, egoism, self-conceitedness. It means being armed with doubt to set forth in search of the truth, proficiency, and the diversity of life.

We should point out that one can observe this state derived from the fourth facet of fuzzy logic not only in the epos "Dede Korkut", but also in other literary texts (in our books about the Oghuz, our tales, eposes, modern poetry…).

To witness once again that such a mode of expression is a normal state not only for our ancient eposes, in general, for our literary thinking, let us consider two lines at the beginning of the famous poem "Mother" by the outstanding Azerbaijani playwright Jafar Jabbarli:

Əgər bütün bəşəriyyət ədüvvi-canım ola,
Ürək süqut eləməz aldığı mətanətdən...
[If the entire humanity is hostile to my soul,
The heart will not collapse from the steadfastness it gets...]

The scheme is the same: *"If..., then..."*. It is possible to restore this scheme in the text as in standard variant. The restored variant would be as follows: "*Əgər bütün bəşəriyyət ədüvvi-canım olsa, (hətta) o zaman (belə) ürək süqut eləməz aldığı mətanətdən...*" [*"If the entire humanity is hostile to my soul, even then the heart will not collapse from the steadfastness it receives..."*]

Through this example, we just wanted to show simply, the fuzzy mindware, in other words, the effort to get alienated from categoricalness not only in our ancient texts, even in our modern literature, as well as in our life is like an oxygen surrounding us, without which we simply cannot survive. Quintessence is defined through this very scheme. The additions, which simply land on it like a dust or spot, are late arrivals which one can easily get rid of. One cannot but feel the precise steps of fuzzy logic in the intellectually–ponderable poems of the outstanding figures of Azerbaijani literature—Fuzuli, Vagif, Seyid Azim, Huseyn Javid, Rasul Rza, Samad Vurghun, Ali Karim, Jabir Novruz, Mammad Araz, Bakhtiyar Vahabzadeh, Nariman Hasanzadeh, Alakbar Salahzadeh, Musa Yagub, Ramiz Rovshan, Vagif Samadoghlu, Rustam Behrudi, Eldar Bakhish. As stated by Professor L. Zadeh, indeed, fuzzy logic is actually a mode of expression of preciseness, punctuality, and these factors appear more vividly and wonderfully in pieces of poetry. With a deep and purposeful attention, we can notice that our entire literature is full of description of the mindware manifesting itself through the scheme of the fourth facet of fuzzy logic and its derivatives.

The study of other pieces of our literature in terms of fuzziness in near future will lead to a considerable broadening of our philological vision, we would say.

In Lieu of Conclusion

We have some significant reasons for considering the text of "The Book of Dede Korkut" from the perspective of fuzzy logic. The first and foremost reason is that the founder of the theory of fuzzy logic, Azerbaijani-American scientist L. Zadeh wished that the principles of the theory of fuzzy logic be initially applied to social sciences, in particular, to the field of philology. We think this book will finally fulfil L. Zadeh's wish. Dear Ustad would be very delighted with it.

Another reason is tracing the formation of our ancient ancestor's mindware in the text of "The Book of Dede Korkut", and seeking an answer to the question how much inclination is there towards tolerance and democratic approach in this mindware. With this in mind, we tried to untangle the clue of deep and hidden relations (both between human beings and linguistic units) within this Epos which is our monumental masterpiece. Geared to it, consequently, it became possible to reveal the democratic mindware pertinent to the creators of the Epos and naturally the bearers of the language and logic of the Epos. All through the study, we noticed that the language of the Epos has an irreversible trace of the democratic processes ongoing in the consciousness of our ancient ancestor. The presence of the factor of a wider choice, the effort to see the (ontological) variety in the world more adequately are expecting their manifestation lurking in the deep layers of consciousness.

There is one more reason which is as if the potentials of applying the principles of fuzzy logic to any literary text as a whole, have gone through "a test" (through practice) in this book. We are convinced that this side of the issue is very significant for the research to be carried out in the humanitarian-philological field in future [1]. The application of the principles of fuzzy logic to any piece from our tales, eposes, folklore, even our modern literature will lead to the proof of the objective and natural presence of democratic shades in the mindware of both our ancient ancestor and our contemporaries. For the time being, we commenced with "The Book of Dede Korkut". We hope that the existence of the democratic mindware, which is an ideological basis of fuzzy logic, in our Epos has been quite vividly reflected in this book and played the part of a beacon for future studies.

K. Abdulla and R. A. Aliev, *The Book of Dede Korkut and Fuzzy Logic*, Studies in Fuzziness and Soft Computing 438, https://doi.org/10.1007/978-3-032-05906-2

The main goal was also to reveal the democratic nature and tolerant spirit of the author's covert mindware, irrespective of what he/she was dealing with (!).

This is a very valuable outcome that there are democratic foundations of the world outlook in our ancestor's mind. This ancient Epos of ours has maintained it through its example. Our ancestor was not contended with Aristotle's logic "either… or…", in other words, with the non-democratic attitude which does not provide a wide choice, but introduced a wider choice in relation to the surrounding environment, in other words, "the exception of the excepting a third", the factor of epistemological recognition, and the elements of doubt and hesitations to thinking as a whole, in accordance with the principles of fuzzy logic, as it was discovered later by L. Zadeh. Thus, he/she displayed the sprouts of democratic approach and the spirit of tolerance in literary works.

As it is obvious, as a literary-artistic asset created through the time by our ancestors since the most ancient, prehistoric period, the epos "The Book of Dede Korkut" hides new perspectives in its deep, covert layers. The familiarization with the mindware of the creators of this Epos from the perspective of fuzzy logic prompts us also that for our ancient ancestors the way of democratic approach was not a coincidental and temporary case in their relation towards the surrounding environment (towards the set comprising their fellow-cavemen and to the system of various sets in the environment). The wide choice, which is one of the main ideological principles of fuzzy logic, and the mode of democratic approach (tolerance) have been fossilized in different expressions in our language up to present as a manifestation of a literary asset pertinent to our people through times, and have maintained the leading line of our ancient ancestor's mode of life up to present.

The co-authors would also like to point out that they are preparing some extensive research titled "The Epos "The Book of Dede Korkut" and Fuzzy Logic" based on this book which serves as an initial approach. Perhaps, in the near future, this research will be available for the readers. Like the present book, this study will also be actually dedicated to the memory of Lotfi Zadeh, the founder of the fuzzy logic, outstanding Azerbaijani-American scientist.

We hope that this book will be interesting for the scholars studying Dede Korkut, as well as artificial intelligence, information theorists, logicians **to the same extent** (!). Now it is high time for information specialists and the scholars working with the texts to understand each other independently and adequately, and get together so that their efforts to reach the goals in the same line yield their real outcome.

Reference

1. Kazimoglu, M.: Fuzziness in the characters of the epos. J. No. 525 (2023)

K. Abdulla and R. A. Aliev, *The Book of Dede Korkut and Fuzzy Logic*, Studies in Fuzziness and Soft Computing 438, https://doi.org/10.1007/978-3-032-05906-2

MIX
Papier aus verantwortungsvollen Quellen
Paper from responsible sources
FSC® C105338
FSC
www.fsc.org

If you have any concerns about our products,
you can contact us on
ProductSafety@springernature.com

In case Publisher is established outside the EU,
the EU authorized representative is:
Springer Nature Customer Service Center GmbH
Europaplatz 3, 69115 Heidelberg, Germany

Printed by Libri Plureos GmbH
in Hamburg, Germany